To Janis
My Best Wishes
Always –

Wm J. Bunton
Sealab Aquanaut
1965-69

Death of an Aquanaut

BY

WILLIAM J. BUNTON

BEST PUBLISHING COMPANY

Edited by Andy Nordhoff.
Layout by Dawña Argenbright.

Cover: Artist's depiction of photograph appearing on page 14 by K.M. Hendricks.

Printed in Hong Kong

ISBN: 0-941332-81-0
Library of Congress Card Catalog Number: 99-66260

Best Publishing Company
2355 North Steves Boulevard
P.O. Box 30100
Flagstaff, Arizona 86003-0100, USA
Fax: 520.526.0370

A Dedication

As harshly unsettling as this historical narration will read to some, after 30 years of suppressed ambivalence, the writing of this sensitive tragedy is in part an overdue reconciliation of the truth based on the facts as I experienced them. Of equal importance, it also is a belated though much-deserved tribute to my long-ago comrades — those now forgotten and never acclaimed "Sealab" aquanauts of the 1960s. It was they, willing to risk their lives to achieve a visionary goal who quite suddenly, had to additionally confront the ultimate humiliation of failure in a misguided though valiant effort to inhabit the unknown hostilities of the deeper ocean depths....

—William J. Bunton

About the Author

Bill Bunton, born 21 July 1933 in the small steel mill/coal mining town of East Vandergrift, Pennsylvania, was raised and schooled primarily in Detroit, Michigan.

Enlisting in the United States Army Paratroops in 1950 during the Korean War, he served 22 months in Korea and Japan with the 187th Airborne Regimental Combat Team prior to being honorably discharged in 1953.

In 1958, Bunton migrated to San Diego, California, and actively pursued a career as a professional deep-sea diver, which eventually led to the position of Diving Supervisor of the U.S. Naval Undersea Warfare Center. During the ensuing years, he further advanced his career by pursuing experimental diving, underwater photography, teaching, writing, and equipment development.

Wanting to specialize in the Navy's beginning involvement in deep mix-gas "saturation diving," in 1965 he was selected to participate in the now historical SEALAB II and III projects. A graduate of the Navy School of Deep-sea Divers as both a 1st Class Diver and Helium-Oxygen Diving Officer, Bunton was one of only a few civilians among the first group of navy divers to be certified a "MAN-IN-THE-SEA AQUANAUT."

In 1969, after the sudden collapse of SEALAB III, Bunton resigned his government position and found himself in the commercial offshore oil industry, diving and supervising jobs off the coasts of California, the Gulf of Mexico, West Africa, Norway, and Japan.

In 1973, he formed a corporation in Hong Kong and for the next six years, surreptitiously attempted to obtain an exclusive salvage permit from the People's Republic of China. His goal was to find and salvage the World War II Japanese-designated "hospital" ship, the *Awa Maru.* Loaded illegally with thousands of tons of strategic metals, plus reportedly carrying a precious cargo of gold and diamonds, the vessel was sunk in the Taiwan Strait by an American submarine supposedly by "accident." However, Bunton had one near insurmountable problem: The *Awa Maru* lay within the territorial waters of communist China. His secret dealings with the People's Republic of China seemingly on the verge of success, a "leak" to the news media triggered worldwide headlines and caused a profound — and to this day mysterious — end to the project.

Based on his unique experiences, briefly described above, Bunton and co-writer Mary Heglar wrote the as-yet unpublished *Target: The Awa Maru.*

Author's Note

Although now rewritten and expanded in detail never before revealed, the basis of *Death of an Aquanaut* is an excerpted segment from the as-yet unpublished book *Target: The Awa Maru* (written by Bill Bunton and Mary Heglar).

But before commencing this serious undertaking, which necessitated reliving my 1965-69 experiences as one of the participating aquanauts during Sealab II and particularly those years as they transformed into Sealab III, striving for ingenuous accuracy within a mist of emotional memories became paramount. Subsequently, once I decided to proceed and to the best of my ability, sincere emphasis was taken in ascertaining all stated withing this ... *memoir* of accounting.

As the more discerning readers will quickly discover, the narrative style of writing in the third-person was used throughout. This method, perhaps grammatically unusual, was nevertheless chosen to avoid the egotistical connotation (in my opinion) of the pronoun "I," both for its otherwise repetitious usage — and certainly the tedious reading thereof.

OTHER PUBLICATIONS BY WILLIAM J. BUNTON:

Problems of Deep Underwater Photography, published by the Marine Technology Society, Washington D.C.

The Offshore Circular Array, a U.S. government research report published by NEL — #1261.

In Pursuit of the Awa Maru, (fiction, co-written with W. Joe Innis), published by Bantam Books, 1980.

Table of Contents

Conquering Inner Space

23 AUGUST 1965 ... OFF SAN DIEGO, CALIFORNIA

At 60 feet, Bill Bunton momentarily halted his descent and urinated. The warmth welled around his thighs until it was slowed just above the knees by the constriction of his wet suit, then spread upward to envelop the rest of his body, up to his neck. Almost instantly he felt comfortable again. Regrettably, the charge would last only about five minutes before it cooled to the temperature of the surrounding water.

Bunton had expected the thermocline — the layer of cold water that separates the oxygen-rich surface water from the heavier bottom strata — but the sudden temperature drop always shocked nonetheless. As was his custom when anticipating the coldness, an hour prior to starting his dive he had gulped down nearly a full quart of water. Now, wanting to ration what was left in his bladder, he forced himself to stop the flow. From his many years of experience and close to 2,000 working dives, he knew he would later need all the warmth he could get.

Above him on the surface where the *Berkone*, the support barge for Sealab II, gently rocked in her five-point mooring, the water temperature was 70° Fahrenheit. The air on this gorgeous late-summer day was 74° F. On the beaches of La Jolla, less than a mile eastward from the barge, crowded the brown bodies and sun-bleached hair that signified Southern California.

August sun warmed the ocean water to a considerable depth. During the first 60 feet of his dive, Bunton had hardly noticed any change in the temperature, maybe three degrees. Then came the thermocline, and within a few feet the temperature dropped another 12° F to about 55, Bunton guessed. Cold. However, not as cold as it would be at the bottom. At 205 feet, the Pacific Ocean, here on the fringe of the continental shelf, was an almost constant 48° F to 50° F.

He peered into the depths, following the white nylon line that guided his descent as it undulated in the light current, then disappeared some eight feet below him. He pointed his flashlight straight down, but it did little to improve visibility. Its focused beam was diffused and blunted by a murky cloud of microorganisms, millions of living particles suspended in the water, floating passively with the current and with no apparent purpose other than to sustain other, slightly larger particles.

He felt a tug on his tether. He looked up to the shadowy figure of his attached diving buddy, Ricky Grigg, only a few feet above him. The tug was a question, a wordless conveyance that had begun with the very first hard-hat divers nearly a century earlier. In that mode of diving, signals had been conveyed through the air hose-cum-lifeline that connected the diver to the surface. The single tug meant "Everything all right?" As an answering tug, it meant "Yes" or "Okay." Because they were within sight of each other, Bunton merely gave Grigg an exaggerated nod of his head in acknowledgment.

If he'd been diving on his own, he would've preferably held a ten pound sandbag and dropped at twice the standard rate of descent of 60 feet per minute, clearing his ears in three or four continuous, partially suppressed exhalations to equalize the ever-increasing water pressure.

But Sealab or the "Man-in-the-Sea Program," although an Office of Naval Research (ONR) project, was one of five programs under the direct control and supervision of the U.S. Navy's Deep Submergence Systems Project (DSSP), which was charged with developing advanced capabilities for underwater fleet operations. So even if it meant less working time on the bottom or more decompression time if he overstayed the pre-planned 20-minute dive, Bunton wasn't about to start arguing with any of their rules; although many were now considered antiquated throughout the rapidly changing commercial diving industry. This was the opportunity for which he had worked, trained and disciplined himself for five years, and he wasn't going to jeopardize everything for the sake of an expedient sandbag.

Before being chosen as an aquanaut, Bunton's employer and workplace had been (and still was) at the Naval Electronics Laboratory (NEL) in San Diego, only now and for at least the next six weeks, he was temporary assigned to Sealab. At NEL, his position as a full-time diver/experimental test mechanic entailed providing underwater support services as needed; which included during the last three years, specializing in deep-diving photography and camera development as well.

Throughout the Navy's numerous research facilities, plus other government funded oceanic oriented agencies and universities, there had been scores of aspiring candidates vying for the Sealab wet suit Bunton was wearing. Foreseeing the need — and opportunity — to enhance NEL's burgeoning subsea projects (deeper) by incorporating mixed-gas diving technology, then convincing his superiors, Bunton was one of only ten civilian divers accepted to participate in Sealab II.[1]

With the guideline sliding loosely through his hand and somewhat encumbered by a bulky camera attached to a quick-release devise on his weight belt, Bunton continued slowly downward.

---

1 To achieve this personal goal, Bunton needed and initiated a formal request through the Ocean Sciences Department of the Navy Electronics Laboratory (NEL), San Diego, to be selected to participate in Sealab II as an ongoing continuation of his assistance in an *in situ* ecological study of living marine organisms (specifically Gorgonian sea corals). This project was under the direction of Dr. Eric G. Barham, Program Manager, NEL. As a secondary function once accepted as an aquanaut, Bunton was requested by Sealab command to also serve as Team III photographer.

Suddenly, a large shape moved below him, vanishing into the gloom as quickly as it had appeared. Shark! he thought, and his left hand clamped around the descending line. Sharks hadn't been a problem during the preparations for Sealab II. A few dorsal fins had been spotted on the surface, and once or twice a diver had seen one below. But they'd never come very close. That meant nothing, though; from time immemorial, sharks were always frighteningly unpredictable.

The name Pamperin flashed into Bunton's mind, and with it a nightmarish picture. Robert Pamperin, an engineer and former lifeguard, was a local skin diver searching for abalone, who'd been attacked in 1959 only a few miles shoreward and slightly south from this very spot.[2] Bunton seldom dove along the La Jolla coast line without the vivid remembrance of the newspaper accounts of the incident.

The shark was so big, the headline stories said, it appeared half of Pamperin's body was stuffed down its throat. Only the upper torso, arms and head — protruded from the great jaws. His terrified diving companion (Gerald Leher), snorkeling only a few feet away, who witnessed the attack, had estimated the length of the shark — probably a great white — to be near 20 feet. After days of searching, Pamperin, nor any body parts were ever found.

Seconds now passed with interminable slowness. At 90 feet, the darkness below Bunton began to lighten as though he were ascending from the depths and rising toward the sunlight, rather than descending deeper. Good: the flood lamps were working.

Below glowed the white shape of the Sealab habitat. Like an elongated spider, the 57-foot-long steel tube, 12 feet in diameter, squatted on the ocean floor on eight short legs. It was situated, slightly tilted, just 60-some feet from the rim of the Scripps Submarine Canyon, whose near-vertical walls dropped to a depth of 700 feet.

A very dangerous location, and certainly not much of a home for human beings, Bunton thought. Yet for the next 45

2 Daniels, Dwight C. "White Shark Blamed for Death of Swimmer." *San Diego Union-Tribune*, 17 April 1994, section A-1 and A-12.

days, men would be living within this tomb-like cylinder, breathing, working, eating, and sleeping just as though they were on the surface. Bill Bunton would be one of them: United States aquanauts.

Underwater pioneers of the new frontier, that's what the news media called them. And the reporters were right. This was a new frontier. Until just the last few years, man's primary goal had always been upward, reaching for and achieving unimaginable distances.

Ironically, human ability to explore the deep ocean was still measured in the low hundreds of feet. Any depths beyond that were considered extremely hazardous and seldom attempted except as simulated in topside experimental diving chambers.

All that was about to change. Sealab I had successfully based four aquanauts for 11 days at 193 feet in the warm waters off Bermuda. Now, down there below Bunton at 205 feet, Sealab II awaited the United States' deepest sustained venture yet into "inner space." Three ten-man teams would spend 15 days each on the seafloor. Succeed here, and our nation's most significant step to date in the exploration of the oceans would have begun.

Already there were plans for another Sealab at a much greater depth; the Navy was talking about 450 — maybe even 600 feet. If humans could live and work at that depth, almost the entire continental shelf — hundreds of thousands of square miles — could be conquered.

Bunton pushed his thoughts aside. First things first. This was Sealab II. And it was up to him as much as any of the other aquanauts to make it work. In the floodlights, Bunton could clearly see the entire length of the cylinder. One end was lit more brightly than the other. This was where a wire mesh cage eight feet long and six feet wide and high protected the underside entrance to the habitat. The aquanauts called it the shark cage, but technically it was the anti-shark cage.

Bunton's gloved fingers hooked into its steel mesh. With a kick, he flipped from head-first to feet-first, and his fins found the silt bottom. Trying not to hurry, yet knowing the importance of each minute, he unlatched the gate, entered and held it open. As soon as Grigg followed him through, Bunton pulled the gate shut and looked around.

He and Grigg waited a few minutes for visibility to return. He detached the 18-inch safety line that secured the burdensome camera to his belt and gently placed it on the bottom, careful not to disturb the sediment any more than necessary.

Bunton checked his watch 1405, or 2:05 p.m. The descent had taken five minutes, at least one more than it should have because of his hesitation over the shark, if it had been a shark. Probably a seal, he told himself.

Still, Bunton was glad the Navy had built the shark cage. The entrance to the habitat was no more than a circular hatch four feet in diameter located on the underside of Sealab and less than five feet off the ocean floor. There was little danger of water entering because the internal gas mixture for breathing was pressurized at just over seven atmospheres absolute,* equaling the ambient water depth of 205 feet. The real danger lay in the fact that a diver leaving the habitat was blind until his head had followed his feet out of the hatch. Without the cage to protect the entrance, one could never know what he was stepping on below. It could be a shark, a razor-toothed eel or more likely in this area and at this depth, a scorpion fish with a highly venomous dorsal fin.

Bunton flicked on the power to the strobe light and rechecked the aperture and shutter-speed settings. Only a few minutes more and *the man* would be here. The photographic coverage would, hopefully, be good enough to make front-page news. For the continuation of the Man-in-the-Sea program and specifically the deep Sealab experiments, favorable publicity was an important means of gratifying the Washington brass who authorized the project's funding.

Ricky Grigg (a marine biologist and Ph.D. candidate from Scripps Institute of Oceanography) was pointing upward. Bunton followed the direction of his arm. The guideline,

* Absolute denotes the total pressure exerted by the surface atmosphere (14.7 psi) plus .445 psi per foot of sea water; and each 33 feet of depth equals one additional "atmosphere."

normally slack because of the allowance for the changing tides and surface swells, snapped taut. As they watched, it jerked — a sure sign that someone was hauling himself down hand over hand. Four minutes later, Commander Scott Carpenter, one of the seven original "Mercury" astronauts, the fifth American in space and the second to orbit the Earth, was unlatching the gate to the shark cage.

Carpenter waved to Bunton, who'd backed into the far corner of the cage, shutter and flash working rapidly. Time allowed for only ten exposures before Carpenter and his diving buddy, Gunner's Mate First Class Wilbur Eaton, disappeared into the habitat. Everything went as rehearsed. The four bolts that secured a temporary cover to the now pressure-equalized entrance hatch, came off smoothly. The two men removed their fins, ducked through the circular opening and were gone.

The conquest of inner space had begun.

Bunton checked his watch again: They'd already used up 20 minutes of "bottom time," the time elapsed from first entering the water until the moment he and Grigg were to begin their ascent.

With Carpenter and Eaton safely inside the habitat, and their entry photographed, Bunton's job was finished. They left the shark cage to face more than an hour of tedious decompression stops. At 80 feet, the roving bell — the oversized, upside-down barrel shaped apparatus that got them partially out of the water while they were undergoing the stops — was waiting for them. Bunton eased in and waited for Grigg to join him before taking out his mouthpiece. He inhaled deeply. The air had a fresh, warm taste. Pumped down by compressor from the *Berkone*, it was real air, not the cold, filtered helium-oxygen mixture that was in his Mark VI scuba tanks and was also the predominate breathing medium inside the Sealab habitat.

Knowing how human voices would sound distorted in the helium atmosphere of Sealab — similar to the high pitch that emanates from "Donald Duck" — Bunton wondered how Carpenter and Eaton, followed by the rest of their team presently descending in pairs, would adjust. Though it would be difficult to understand each other at first, they'd have to adapt. Ten men couldn't spend 24 hours a day with each other for 15 days without talking. Correction, Bunton reminded himself. It was 15 days for nine men, but 30 days for Scott Carpenter. He was scheduled to remain in the habitat as leader of not only Team One but also Team Two. Bunton himself was to go down with Team Three for the final 15 days of the 45-day experiment.

The roving bell broke through the surface. Bunton and Grigg ducked out from its underside and swam the few yards to the side of the *Berkone.* Waiting to be helped aboard, Bunton noted the time: It was 3:30 p.m.. The ascent had taken 70 minutes, with decompression stops every ten feet of the final 80 feet. It felt good to be on the surface again, freed from the constraints of the bell.

The *Berkone's* fantail was almost deserted. A few crewmen, dive-master Dan Price and a couple of private contractors. Where had all the media gone? Ninety minutes ago, when Bunton and Grigg had commenced their dive, there had been at least 20 of them — newspaper reporters, TV people, magazine writers, photographers. Now, there wasn't a camera or notebook in sight.

Then Bunton remembered the command van; which in fact was not a part of the *Berkone* superstructure, but a large compact trailer minus wheels, and hoisted aboard earlier by crane. It's function for the next 45 days, would serve as Sealab's on scene headquarters. That's where the media were obviously congregated — packed in and around like sardines.

Carpenter and his nine-man team would all be down in the Sealab capsule by this time and in verbal contact with the *Berkone.* A loudspeaker, hooked up to the two-way telephonic system, had been attached to the outside of the van; Carpenter's voice, could be heard "quacking" throughout the barge.

While Bunton was helping dive-master Price and Grigg clean and stow the diving gear, Captain George Bond, who also served as chief medical officer, suddenly appeared on the fantail. Bunton greeted the man, and Grigg threw out an irreverent "Hi, Papa."

To the aquanauts, Bond was "Papa Topside," a term embodying both affection and respect. It was typical of Bond that he had taken the time, even during this crucial first hour commencement of Sealab II, to check on his two divers. He always made a point of letting his men know that he was concerned. The aquanauts knew his solicitude was genuine.

Bond never asked anyone to do anything he hadn't already done himself. Although past middle age and somewhat overweight, he had made one of the exploratory dives with Carpenter and Bunton to the ocean floor earlier in the year when they were searching for the best site for Sealab II. Because the helium-oxygen mixture hadn't arrived from Panama City, Florida in time, the 205-foot dives had all been made on compressed air. Bond had no business trying it. Due to the insidious effects of nitrogen narcosis (commonly referred to as "rapture-of-the-deep"), air dives to that depth were not only demanding for even a younger diver, they were extremely risky. Bond knew that. His men knew it too, and they appreciated his gesture.

Assured that Bunton and Grigg were suffering no ill effects from their 20 minutes on the bottom and subsequent decompression, Bond excused himself. Thank God for Papa Topside, Bunton thought. Though not officially designated Project Director, he was without a doubt as on-site Deputy for Medical Affairs and indisputably in charge of Topside Control, the glue that held the Sealab diving teams — two officers, 16 enlisted men and 10 civilians — together.[3]

To some of the Navy brass, civilian divers such as Grigg and Bunton were an unwelcome intrusion. They were open links in the chain of command, a subtle nuisance that had to be tolerated at the moment.

---

[3] **Team I:** M. Scott Carpenter, CDR, USN (also Team II); Robert E. Sonnenburg, LT, MC, USNR (also Team III); Berry L. Cannon (Navy Mine Defense Laboratory); Thomas A. Clark (Scripps Institute of Oceanography); Billie L. Coffman, Torpedoman's Mate First Class, USN; Wilbur H. Eaton, Gunner's Mate First Class, USN; Frederick J. Johler, Chief Engineman, USN; Earl "A" Murray (Scripps Institute of Oceanography); Cycril J. Tuckfield, Chief Engineman, USN; Jay D. Skidmore, Chief Photographer's Mate, USN. **Team II:** Robert A. Barth, Chief Quartermaster, USN; Howard L. Buckner, Chief Steelworker, USN; Kenneth J. Conda,

Continued Next Page

He had seen it happen before — the friction, the coexistence by sufferance, the gradual breakdown in communications that seemed to nag many research projects in which the military was forced to work with civilians; at least with civilians who were considered experts in their field. On some jobs, that might work. But Bunton hoped that due to the extreme seriousness of what was at stake, their individual skills would be a recognized contribution — and fully utilized....

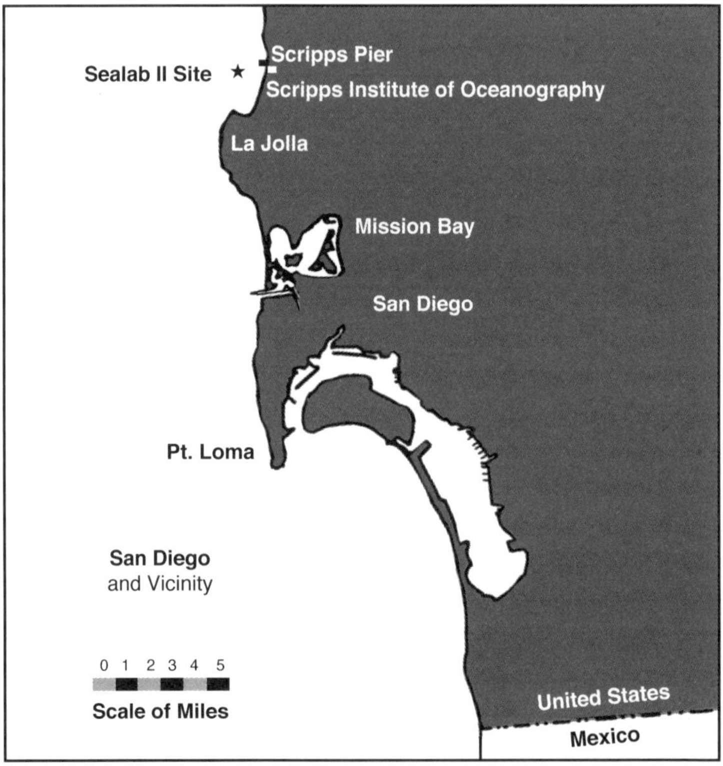

Map of Southern California coast showing Sealab II site 1,000 yards west off the end of Scripps Pier.

Torpedoman's Mate First Class, USN; George B. Dowling (Navy Mine Defense Laboratory); Arthur O. Flechsig (Scripps Institute of Oceanography); John F. Reaves, Photographer's Mate First Class, USN; William H. Tolbert (Navy Mine Defense Laboratory); Glen L. Iley, Chief Hospital Corpsman, USN; Wallace T. Jenkins (Navy Mine Defense Laboratory). **Team III:** Robert C. Sheats, Master Chief Torpedoman's Mate, USN; William J. Bunton (Navy Electronics Laboratory); Charles M. Coggeshall, Chief Gunner's Mate, USN; Richard Grigg (Scripps Institute of Oceanography); John J. Lyons, Engineman First Class, USN; William D. Meeks, Boatswain's Mate First Class, USN; Lavern R. Meisky, Chief Shipfitter, USN; John M. Wells (Scripps Institute of Oceanography); Paul A. Wells, Chief Mineman, USN. [ONR Report ACR-124, "Project Sealab Report", 8 March 1967]

23 July 1965: *The Sealab II christening ceremony at the Long Beach Naval Shipyard, with all 28 aquanauts in attendance.*

(USN Photo)

23 July 1965:
Just prior to the christening ceremony of Sealab II at the Long Beach Naval Shipyard, Team III civilian aquanauts (left to right) J. Morgan Wells and Ricky Grigg — both from Scripps Institute of Oceanography. Note the anti-shark cage that protects the entrance area to the habitat.

(Photo by Bill Bunton)

26 August 1965: *1,000 yards west of the Scripps Institute of Oceanography pier, Sealab II being prepared to descend 205 feet to the sea floor, with the Berkone in support.*

(USN Photo)

August 1965: *Aboard the Berkone, Dan Reeves, topside Master Diver USN, inspects the PTC (Personnel Transfer Capsule).*

(USN Photo)

2 October 1965: *Aquanaut Richard Grigg making an excursion dive from the Sealab II habitat to the rim of Scripps Canyon at a depth of 250 feet.*

(Photo by Bill Bunton)

August 1965: *Aboard the Berkone, aquanaut Bill Bunton, who initiated the development of this prototype 400-exposure, 70 mm U/W camera (manufactured by Hydro Products Corp.).*

(Photo by Don Dornan)

12 October 1965: *Aquanauts CDR Scott Carpenter, USN and Bill Bunton during a press conference held at Scripps Institute of Oceanography. Aquanaut Billie Coffman, Torpedoman First Class USN, is in center background.*

(USN Photo)

27 September 1965: *Aquanaut and Team III Leader Robert Sheats, Master Chief Torpedoman (DV) USN, providing dive/work schedules to three attentive aquanauts. Note: Although seven Team III aquanauts formally commenced their 14 days on the bottom 26 September (until approximately 10-11 a.m. 10 October), Sheats, Bunton, and William Meeks, Boatswains Mate First Class (DV) USN, descended to the habitat 28 hours earlier (10 a.m. 25 September), joining Team II during their last day and night prior to ascending the following morning.*

(Photo By Bill Bunton)

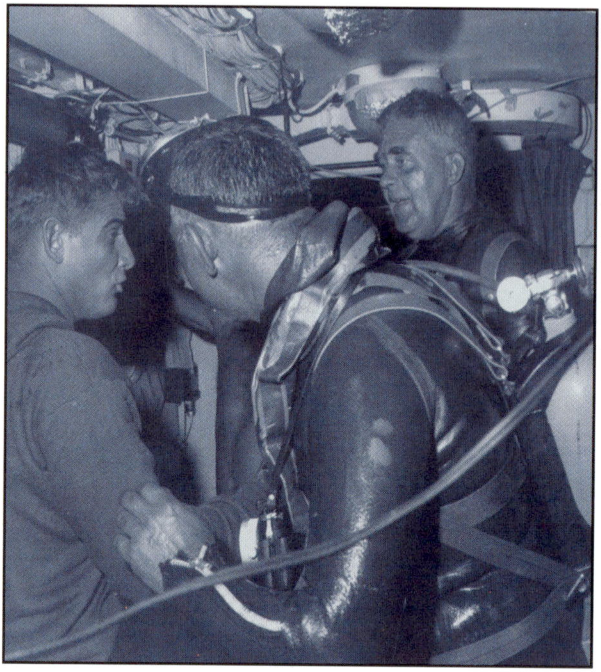

September 1965: *Captain Bond and Captain Mazzone make a brief 205-foot mixed-gas scuba dive to visit Sealab and Team II. From left to right: Howard Buckner, Chief Steelworker (DV) USN, Mazzone, and Bond.*

(USN Photo)

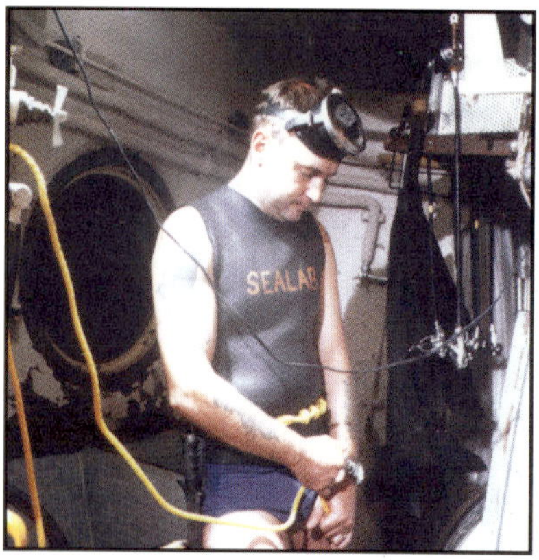

26 September 1965: *Aquanaut Robert Barth, Chief Quartermaster (DV) USN (Team II), after 14 days on the bottom (since 12 September) preparing to and then vacating Sealab II for the last time, making a short breath-holding 30-foot dive to the PTC (Personnel Transfer Capsule).*

(Photos by Bill Bunton)

5 October 1965:
(Above and at right) Aquanaut and Team III Leader Bob Sheats, after descending down the near vertical wall (with sporadic rock outcroppings) of Scripps Canyon to a record depth of 300 feet.

(Photos by Bill Bunton)

28 September 1965:
(Left) Aquanaut William "Billy" Meeks, Team III, making an excursion dive into Scripps Canyon, pauses at a rock outcropping to examine a forming gorgonian sea coral at a depth of 266 feet.

(Photo by Bill Bunton)

October 1965: *Aquanaut Bob Sheats returning to Sealab II after a long excursion dive. Note the four-foot diameter entrance hatch.*

(Photo by Bill Bunton)

October 1965: *Same dive and near-freezing diver hurrying to strip off his wet suit to take a hot shower (in background).*

(Photo by Bill Bunton)

12 October 1965: *Nine Team III aquanauts after undergoing 30 hours of decompression, answering questions at a press conference held at Scripps Institute of Oceanography. Project Director, Captain L. Melson, is at the podium. Note: Team Leader Bob Sheats is not available due to suffering a mild case of the "bends," which required 12 additional hours of decompression in the DDC.*

(USN Photo)

12 October 1965: *Captain George Bond, MC (Sealab Topside Control and Deputy for Medical Affairs), speaking before a press conference held at Scripps Institute of Oceanography.*

(USN Photo)

October 1965: *Team III aquanaut Charles Coggeshall, Chief Gunner's Mate (DV) USN, preparing for an excursion dive, scrupulously setting up his Mark-VI diving unit. Note the TV monitoring camera outside porthole.*

(Photo by Bill Bunton)

October 1965: *Team III aquanauts (left to right) John "Shorty" Lyons, Engineman First Class (DV) USN, and LT Robert Sonnenburg, MC, USNR. Doctor Sonnenburg also was a member of Team I.*

(Photo by Bill Bunton)

25 August 1965: *Sealab II being prepared for its next day descent to the sea floor 205 feet below.*

(Photo by Bill Bunton)

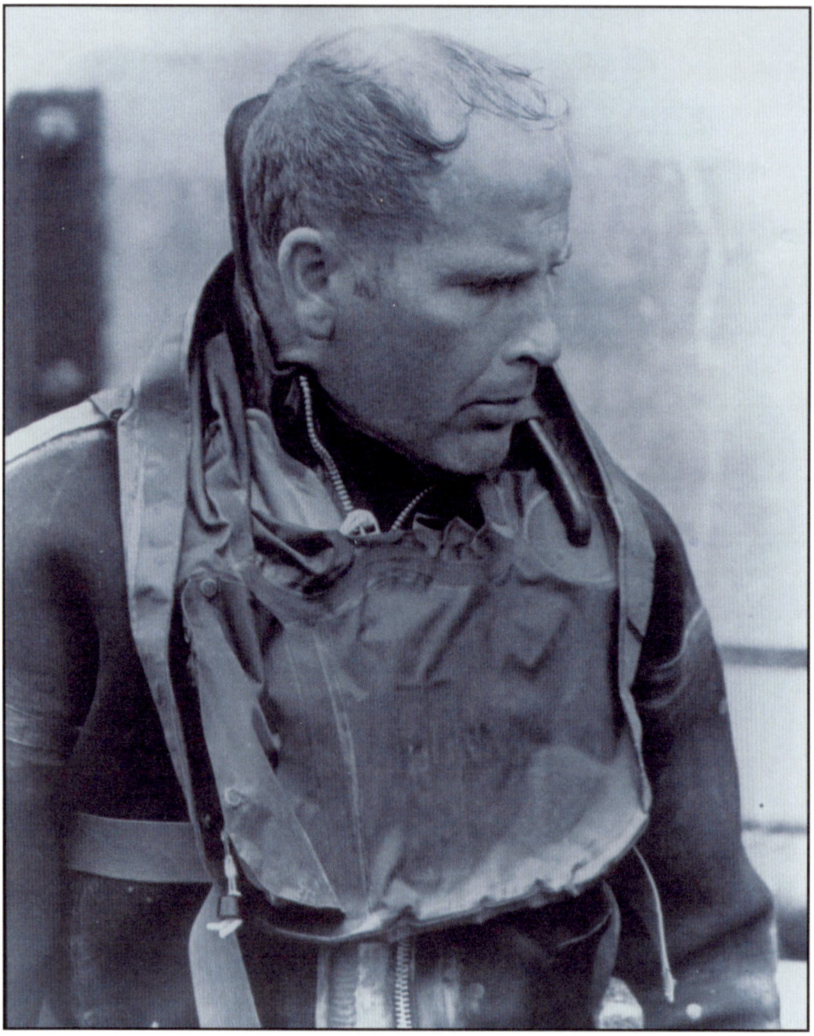

28 August 1965: *Aquanaut Bill Bunton after a 20-minute, 205-foot decompression dive to photograph the first two-man entry of Team I (aquanauts Team Leader CDR Scott Carpenter and Wilbur Eton, Gunner's Mate First Class (DV) USN), into Sealab II.*

(Photo by Don Dornan)

5 October 1965: *Team III aquanaut Bill Bunton returning to Sealab II after he and Robert Sheats make the first 300-foot descent in to Scripps Canyon.*

(USN Photo)

12 September 1965: *Lifted aboard the Berkone, the PTC preparing to mate with the DDC (Deck Decompression Chamber), where the nine aquanauts from Team I will undergo approximately 30 hours of decompression.*

(USN Photo)

12 September 1965: *After 15 days on the bottom, the PTC surfaces carrying nine aquanauts from Team I. Note: CDR Carpenter remains in the habitat an additional 14 days as leader of Team II.*

(USN Photo)

August 1965: *The "Roving Bell" being lowered into the water, which was utilized to control decompression stops for all surface dives to the Sealab II site.*

(Photo by Bill Bunton)

24 August 1965: *Less than one mile off La Jolla, California, Sealab II awaits its descent in 48 hours.*

(Photo by Bill Bunton)

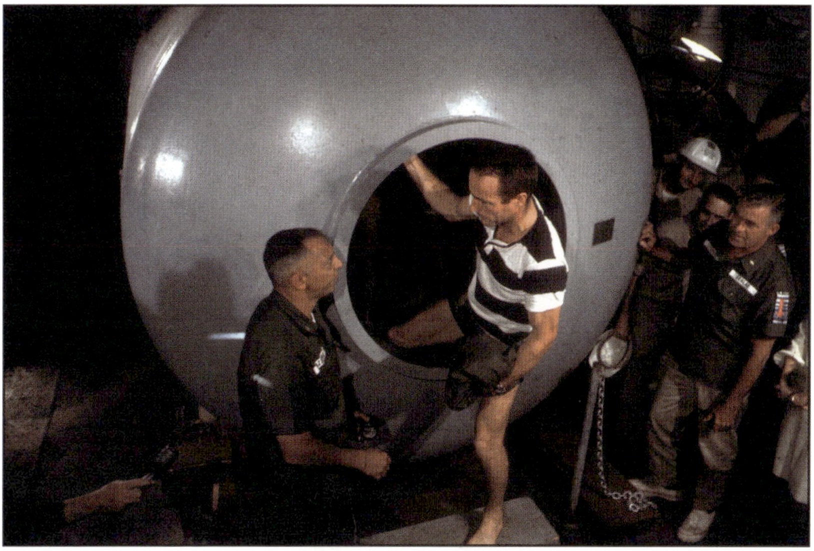

28 September 1965: *Greeting CDR Scott Carpenter as he exits the DDC after spending one month in Sealab II (plus 30 hours in decompression). From left to right, Captain Walter Mazzone, Physiological Control/Diving Operations Officer, Scott Carpenter, and Captain George Bond to his rear.*

(USN Photo)

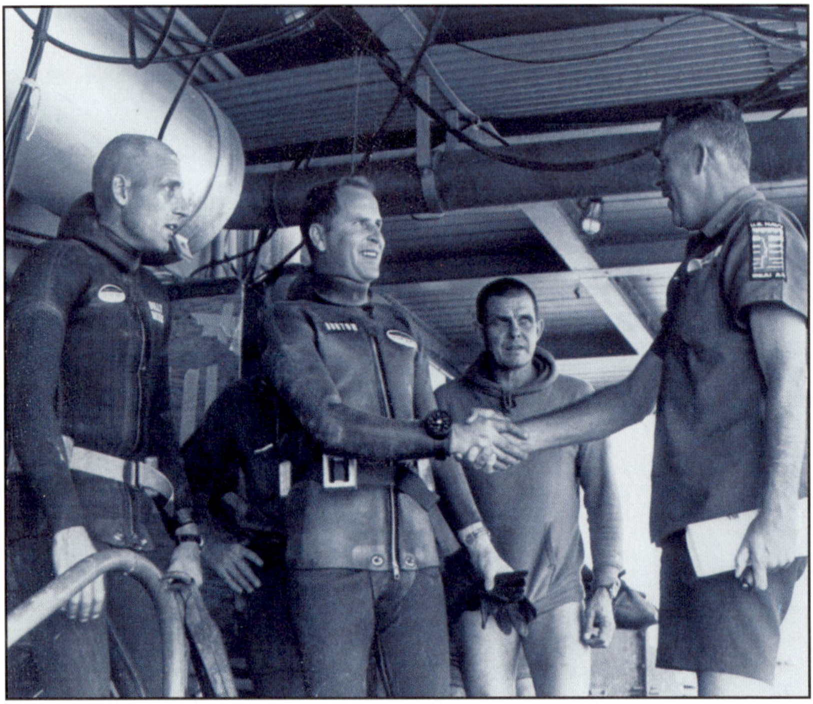

25 September 1965: *Aboard the Berkone and only minutes before their descent to Sealab II, Team III aquanauts "Billy" Meeks and Bill Bunton have a few final words with "Papa Topside." From left to right, Meeks, Bunton, Team I aquanaut Cyril Tuckfield, Chief Engineman (DV) USN, and Captain George Bond.*

(Photo by Don Dornan)

The Next Step

15 AND 16 FEBRUARY 1969 … OFF SAN CLEMENTE ISLAND

The crane was one of the largest in the world. During World War II, the Nazis had built three of these floating monsters in the shipyards of Kiel and Hamburg to clear the entrances to their naval bases. This crane was also one of the few pieces of seagoing hardware the Allies had the foresight to spare in the wave of ship-scuttling and factory-dynamiting that occurred in the months following the war's end.

Still functional 24 years later, this relic from a violent past was still the best of its kind, and it was serving a new kind of conquest: The conquest of inner space. Eight thousand miles from its birthplace by the stormy North Sea, it now rode the swells of the Pacific, 70 miles off the coast of San Diego.

Today, the Pacific was far from its usually placid self. Winds gusted to 35 miles per hour out of the south, unusual for this time of year. With the normal prevailing northwesterlies, the crane would have been in the lee northeast of San Clemente Island, less than a thousand yards away. As it was, a heavy chop battered the crane's platform, and the steel giant strained against the chains of its five-point mooring. The articulated boom, towering nearly seven stories above the surface, groaned in its joints.

"Is it swaying?" asked Wally Jenkins.

Standing beside Jenkins, Bunton focused on the rust-colored skeleton of the boom against the background of dark, hurrying clouds. The pitching sensations could've come from the crane itself, the clouds or their support barge being pounded by the waves.

"Can't tell," Bunton answered. His gaze followed the thick cables running from the tip of the crane down more than 100 feet, where they were attached by brawny shackles to rings welded to a vast steel cylinder bobbing in the water. The veteran aquanaut studied its structural configuration, which during the past two years had been repainted from white to yellow, modified and lengthened to 63-feet. "Know what it looks like? Like a damn truck, a gas tanker."

Sealab I had put four men on the ocean floor 193 feet down for 11 days at a cost of $148,000 in July 1964. In the fall of 1965, Sealab II saw three ten-man teams stay 15 days each at a depth of 205 feet at a cost of $1,400,000. Now, in February 1969, Sealab III was about to put five nine-man teams, for 12 days each, 610 feet below the surface at the staggering cost of $10,000,000 — almost a sevenfold increase in just five years.

Bunton and Jenkins continued watching as the habitat was prepared for lowering. On its bow, the words "U.S. NAVY SEALAB" painted two feet high, were visible intermittently between breaking waves. Bunton fixed on the name as if in a trance, when suddenly he realized that the bottom lettering was no longer above water. He looked at the crane. The gigantic drum holding the steel cables was revolving ever so slowly.

Sealab III was going down. Another great adventure was about to begin.

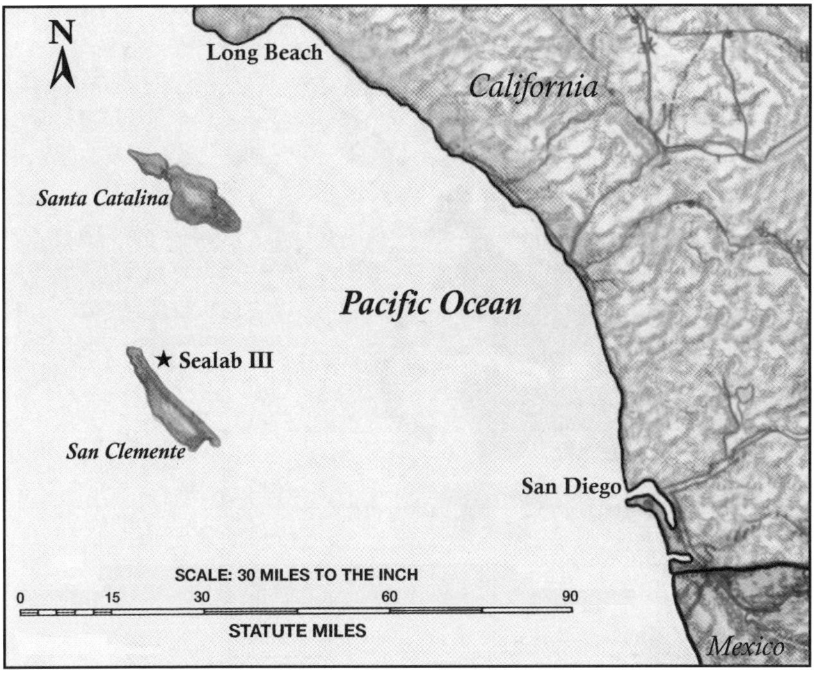

Map showing location of Sealab III off San Clemente Island — just south of "Wilson Cove."

Beginning of the End

Like so many others, Bunton was caught up in the euphoria that followed the overwhelming success of Sealab II. The theory of saturation diving* had been proven beyond challenge. Humans now could not only work — but actually live 200 feet beneath the surface of the sea. For the U.S. Navy and its aquanauts, Sealab II had been glory.

It meant much more than that to private industry, especially off-shore oil and related companies. To a large extent through the early 1960s and certainly for prior decades, they — both foreign and domestic — depended on the U.S. Navy's deep water diving technology, that would eventually make the entire Continental Shelf accessible for exploration.

---

* Saturation diving is the technique whereby and within 18-24 hours of compression at the planned depth of the dive, all tissues of the body become *saturated* with the inert gas helium (substituted for nitrogen found in air, which, when breathed under pressures exceeding depths greater than 120 feet, increasingly affects one's mental skills similar to a narcotic or drug-induced stupor, and eventually leads to unconsciousness). The revolutionary advantage of this method of diving and whatever the sea-floor task, is not being penalized by repeated hours of decompression after each dive of even short bottom times. But when saturated and that period ends, be it six days, six weeks or six months, the diver is only subjected to one lengthy decompression — based on depth alone — regardless of duration.

Secretary of the Navy Paul Nitze awarded all the aquanauts a Navy Unit Citation in recognition of their tremendous achievements in advancing the conquest of inner-space. Additionally, along with the nine other non-military aquanauts, Bunton was awarded the Navy's top individual honor: The prestigious Superior Civilian Service Award.

His continuation in the project as it evolved into Sealab III, was encouraged by Dr. G.H. Curl, the far-sighted Coordinator of Ocean Sciences at NEL.[4] It was his previous recommendation that had initiated Bunton's acceptance in Sealab II.

Although he would no longer be one of the "down divers" for Sealab III, Scott Carpenter was assigned the responsibility of Deputy On-Scene Commander. He was also designated Special Assistant for Aquanaut Operations. But, many of the old-timers from Sealab II were gone. They were either transferred to other duty stations or for their own personal reasons, decided to leave; which for each individual aquanaut, was strictly a voluntary decision.

However, there was a steady influx of new naval officers assuming key positions within the project. And in addition to the 40 originally selected aquanauts, there were quite unexpectedly, to be five more — including Jacques Cousteau's youngest son, Phillipe. That meant instead of the pre-planned eight divers per team, now nine-man teams would crowd into Sealab III on the ocean floor. For whatever realistic contribution these five were suddenly to make, except perhaps celerity enhancement and/or PR value ... must have been determined at the highest levels in the Navy.

[4] Once again being selected as an aquanaut for participation in Sealab III, and under the direction of Dr. William Cummings of NEL (in a transitional state of being renamed the Naval Undersea Warfare Center (NUWC), a leading authority on bioacoustics, Bunton took on the responsibility of a scientific program important to submarine warfare and sonar detection. The project involved an array of hydraphones and a specially designed time-lapse camera to record, photograph, and identify various fish species that emanated different (and potentially sonar distracting) sounds. And, as during Sealab II, the Navy requested he be designated Team IV photographer. Note: In late 1968, due to the now known and complex time consuming demands of his primary NUWC project, he recommended to CDR Carpenter his teammate Wally Jenkins be assigned those duties, with his assistance training Jenkins in the usage of his PC-770 camera. [Bunton journal entry]

Also, to the dismay of many, the revered Captain Bond, known worldwide as the principal creator of saturation diving and the driving force of the Sealab project since its concept, had been stripped of much of his authority. Nominally, he was still the Chief Medical Officer and Principal Investigator. But, in actuality, Papa Topside now retained little influence over the grave, yet unmistakable pervasion of escalating events.

As the first few months passed and gave way to these stark realities, Bunton was becoming concerned over a gradual change in the Sealab command structure. Topside authority was being taken out of the hands of those who specialized in diving research, who'd proven their qualifications and dedication over many years, and put under the control of regular Navy line officers.

A similar change was taking place at the working level with the divers who would risk their lives to make it work. Few of the newest American aquanauts were professional divers in the truest sense. Many had little or no deep-diving experience in the open seas; not the civilians chosen for their scientific marine programs, nor the Navy personnel with skills more fitted for standard shipboard operations. True, they were recent or even older graduates of the U.S. Naval School, Deep Sea Divers, and certified "Divers First Class" (or its equivalent, in the case of several foreign aquanauts participating), qualified on air to 285 feet and helium-oxygen to 320 feet. But for the most part, their quasi deep-depth "experiences" consisted of closely monitored non-oceanic dives in enclosed pressurized wet chambers (two stories high) during training, making only a few exceeding 150 feet — and usually only one each to their maximum qualification depths. Thereafter, during the ensuing months that stretched into more than two years, this was then followed by infrequent spurts of shallow-water dives in the bays of San Diego, Long Beach, and San Francisco.

One rumor that had become fact was the extreme depth at which Sealab III would take place: 610 feet. Few divers would admit to the fear this figure engendered in them. The depth was nearly three times that of Sealab II — deeper than only a handful of divers had ever gone before, none of whom had stayed down for any extended period of time like the planned 12 days for each team.

When the Navy confirmed that rumor, the first seeds of serious doubt took root in Bunton's mind. He believed that 610 feet in the open sea was just too deep, too soon. Forty-five divers in rotating teams would be doing dozens of sophisticated in-water projects that would, in varying degrees of equipment development, be totally dependent on an overhead support group of 20 or more competing private contractors who appeared to be coming and going without end.

The setup, Bunton felt, begged for failure: Technology outrunning human control, bloated Washington D.C. staffs tripping over each other, other on-scene topside personnel naive about the contemplated depth or full of grandiose plans, and insufficiently trained aquanauts…

Remembering back on that premonitory day when he was requested to finger print his own death certificate, Bunton consciously, now had to ask himself what he no longer could avoid: Was Sealab III becoming too much of a risk — a disaster just waiting to happen?[5]

Somewhat sensing history being made, though mainly due to his obligatory and ongoing reportage to his superiors at what was recently renamed the Naval Undersea Warfare Center or "NUWC" (previously NEL), he'd begun keeping a personal journal as he started Sealab III. By mid-February 1969, it held a litany of foul-ups, postponements and breakdowns.

During the last, seemingly endless months, valves stuck, lines fouled, equipment malfunctioned. A civilian aquanaut nearly drowned in a pool during a training exercise in Long Beach. One of the two personnel transfer capsules (PTCs) fell into the water — with its hatch open. The habitat flooded

5 After his ("preferred") right index finger was inked and printed in Box 21 of "CERTIFICATE OF DEATH" (Form: NAVMED N [rev. 4-58] front), Bunton was advised by Navy medical authority to keep the document safely in his possession, or within his family, in the event positive identification was ever necessary. Note: The exact date this transpired is not known, but possibly 22-23 January 1969 during a final 2-day physical exam performed by Team IV LCDR Mark Bradley, MC, USN at the Long Beach Naval Shipyard. [Bunton journal entry]

during a bayside shallow-water checkout. Cables were miswound onto winch drums. A PTC was tested at 550 feet and came up filled with water. Repairs dragged. Testing procedures and arrivals of new equipment were delayed time and time again.

And so it went. For more than three years after Sealab II, Bunton's journal recorded lengthy periods of delays spiced with short periods of rushing to meet sudden deadlines. Eventually the starting date of 17 February 1969, was finally set, and all hands bent to meet it.

Not quite all hands. The photographer scheduled for Team One, Lieutenant Laurence Bussey, apparently foresaw what was coming and wanted no part of it. He hurriedly put in for a transfer and left San Clemente Island the day of the first preparatory dive on 16 February.

And Scott Carpenter, frustrated and concerned by the plague of problems, considered leaving the project shortly before it began.[6] He stayed on only out of loyalty to the rest of the aquanauts.

Bunton stayed too; even though his pessimistic laced reports of the past few months caused his NUWC department head to urge him "...give it up," and return to his job.[7]

---

6 As conveyed during a private conversation (over coffee) requested by Scott Carpenter, who primarily and informally wanted to know Bunton's personal feelings regarding the events that led up to and caused Berry Cannon's death. Bunton advised him that rather than rehash the problems of Sealab that troubled him, particularly during the past nine months which had already brought to his attention more than once, that now was not the time for further elaboration. Besides, once he returned to NUWC, which he would in a few days after getting all his equipment out of Sealab, his boss will likely want him to write a comprehensive report that will be available to all. But for whatever value his interim thoughts were worth now, he had been told four days earlier by Sealab command to prepare a written statement on that subject, which he was still working on. It would be turned over to the Board of Inquiry hearings set to convene 28 February (which, of course, Carpenter also knew); and that he would give him, or see that he received a copy of it when finalized plus his final NUWC report if/when written. This meeting took place 24 February 1969, A.M., at either Sealab CIB (Command Information Bureau) or "Mercer Airline" terminal coffee shop, both located within the Long Beach Airport complex; or possibly the "Outrigger Motel," where Bunton and Carpenter frequently stayed during the Sealab II and III years of 1965 through early 1969. Note: (1) Carpenter would never receive copies of either said statement or potential report. See "Footnotes 14" and "13" for reasons; (2) Mercer Airline had the exclusive contract to provide all flights to and from San Clemente Island, for Sealab III personnel, associated contractors and the media. [Bunton journal entry]

7 Mainly telephonic, and more frequently commencing the latter part of 1968, these update reports concerning the repeated delays and serious, seemingly endless problems

Continued Next Page

But regardless of his growing apprehension, he also felt an obligation of commitment and just could not bring himself to quit and walk away. Not after almost four years of effort devoted to what the project was attempting to achieve; which, for his own personal reasons, considered its success crucial to his post-Sealab III plans of expanding NUWC's research diving and service capabilities.

So despite all the ominous signs, he was a part of it and emotionally continued to rationalize that — the decision-makers surely knew what they were doing. If only they all could hang on a few more weeks, it would all be over....

The spot chosen for the placement of Sealab III was ideal. West-northwest of San Diego, the water was deep yet clear. Nearby San Clemente Island provided shelter for surface vessels and living quarters for the support personnel. The Navy already had existing facilities on the island, which lay in the vast top-secret military sea range used for testing the Polaris and Poseidon missiles, for surface-ship bombardments and for various underwater experiments.

Everything was bigger and better for Sealab III. The *Berkone*, the earlier support barge, had been replaced by the *Elk River* (IX-501), a 203-foot former rocket-launching ship especially modified and outfitted for deep-sea diving projects. Gone, too, were the old roving bells; at the more extreme depths of Sealab III, there could be no free dives to the bottom. Aquanauts would ride to the ocean floor in the supposed comfort of enclosed, heated PTCs (personnel transfer capsules), after already being compressed to bottom depth in two large pressurized deck decompression chambers, or DDCs, aboard the *Elk River.*

When the Sealab III habitat touched bottom at 8:55 p.m. on 15 February, Bunton was back in his quarters on the island. He needed all the sleep he could get. The previous 48 hours of practically non-stop preparations had totally exhausted him, which included two separate 60 minute working dives prior to

of Sealab III, were conveyed directly to his immediate supervisor, Art Schlosser, Hd. Ocean Eng. Div. NUWC, San Diego, or other concerned personnel within his divisional control. [Bunton journal entry]

Sealab's beginning descent. The last entry he recorded in his journal the night of 14 February was: "Everything rush — rush. Men tired — little rest. Bad!"

The next day would start with an early-morning briefing, then for him a 4 p.m.–to–midnight watch in the command van helping to monitor the first dive.

The skiff brought Bunton back to the *Elk River* at 6:45 the following morning. Aquanaut Wilbur Eaton greeted him as he boarded the vessel.

"Mornin' Bill. We've got trouble already."

"What kind this time?"

"The habitat's sprung a leak."

"How bad?"

"Dunno. I think they're getting ready to send the first team down to fix it. See?"

Bunton followed Eaton's pointing finger to the foredeck, where the two seven-foot diameter PTCs, those containers that were to be the aquanauts' elevator cages to the bottom, were sitting like giant crock pots. Three or four men were busy checking the helium-oxygen flasks, each nearly six feet high and secured vertically against the outside of the capsules.

"Anybody inside them yet?" Bunton asked.

Eaton shook his head. "Not in the PTCs ... they're all in both DDCs, though."

The airtight deck decompression chambers, each 24-feet in length, were located just below and perpendicular to the deck on the *Elk River.* Their purpose was to compress the divers before their physical descent to the sea floor, by gradually increasing atmospheric pressure to the same pressure as the water's at the depth of the planned dive, and to reverse that process by decompression after the dive. At the same time as the pressure was increasing, the "air" content of oxygen-nitrogen was slowly changing to nearly pure helium, then reversed. (At bottom depth, the oxygen in the breathing medium would be sustained at less than two percent — but its life-sustaining partial pressure was comparable to the approximately 21% found in the atmosphere at sea level.)

Because of the length of compression and decompression time required for dives as deep as Sealab III, the DDCs were more than mere hyperbaric chambers. They were compact but fairly spacious living quarters (capable of being divided into two separate compartments) with most of the comforts of home including TV, a shower and, through a small airlock, hot food. In the bulkhead of each DDC was a separate attachment lock, through which the divers would transfer to the PTC before a dive and return the same way. Before the actual transfer took place, the capsule would also be pressurized to bottom depth.

The briefing began promptly at 7 a.m. Conducting it was Captain William M. Nicholson, who'd recently been appointed overall project manager. His career up until then had centered on naval engineering and ship designing. Nicholson confirmed what Bunton had already heard: Sealab, lowered the previous night, was leaking gas.

The situation was not dangerous. The habitat was still vacant, and the leaks were being controlled by pumping additional gas through the main umbilical at a slight over-pressure to keep the water out. The potential problem was not one of water coming in, which could be dealt with if only a small amount entered, but one of gas escaping. The rate of loss was about 3,000 cubic feet per hour, Nicholson said. At that rate, the gas storage tanks on the *Elk River* and the support supply barge would be depleted in just over a day.

"Are we bringing it up?" someone asked.

Nicholson shook his head. "No. Not yet, anyway. We still have some time to decide. At the moment, the thinking is to send down the entry team and see whether we can find those leaks and plug them."

Bunton's descent to the bottom and 12-day stay wouldn't begin for several weeks. He was scheduled to go with Team Four. Until then, like the rest of the aquanauts, he had to perform various topside support functions.

At 3:45 p.m., he checked in at the command van to relieve Martin Harrell, a fellow civilian diver, who quietly began briefing him. Commander Jack Tomsky, a nonscientist who specialized in ship salvage and repair, was the designated commanding officer for Sealab III. During the day, Harrell related, Tomsky had made the decision to compress four divers from Team One to bottom pressure in three hours — five times as fast as the 15 hours the medical advisors recommended and planned for. It was now crucial that they go down, enter the habitat, and seal the leaks as quickly as possible.

At the same time, the remaining five members of Team One in the other DDC, were being compressed at the regular rate. They would either act as a backup squad or, if all went well, descend as scheduled. The first four had already completed compression in their DDC and were awaiting the go-ahead for transfer to the PTC.

"Barth's team," Bunton confirmed. That team had been picked weeks before for the job of opening and securing Sealab for all who were to follow during the next two months: recently promoted to Warrant Officer — Robert Barth, 38, the only remaining and still participating aquanaut of the U.S. Navy's saturation diving experiments which began in 1957; Berry Cannon, 33, a civilian engineer at the Navy's Mine Defense Laboratory in Panama City, Florida; Richard Blackburn, 30, an Aviation Ordnanceman First Class, the youngest and strongest of the team and the only newcomer to Sealab; and as Lieutenant Bussey's recent replacement, John Reaves, 39, presently a civilian photography specialist who had just retired from the Navy as a Photographer's Mate First Class.

Bunton settled into his seat in front of the control panel. His job was to monitor and log the readouts of a number of gauges indicating such data as differential pressure (DP) being pumped inside the habitat from the *Elk River.*[8]

By 5 p.m., the first team of divers had entered the PTC and completed the check-off procedures. At 5:25 the capsule was lifted off the DDC and lowered into the water. At 6:15 it reached bottom depth, suspended about five feet above the sea floor, and was positioned about 40 feet from Sealab III. At 6:54

[8] Commencing at 4 p.m. through 11:50 p.m., 16 February 1969, all times, events, numbers, quotes and paraphrased comments are either (1) transcribed from Bunton's copy of the official log (Form MDL 8500/1 [8-68]) that was kept in the command van "communication watch"; (2) personal, on-scene observations, including verbal usage by said parties at the time; and/or (3) during subsequent direct testimony witnessed during the 28 February - 12 March 1969 Board of Inquiry hearings. [Bunton journal entries]

two of the four — Barth and Cannon — opened the PTC hatch and dropped into the water. Dragging their breathing-gas umbilicals, they swam toward the habitat. Reaves and Blackburn remained in the capsule, both as a support team and to stay in telephone contact with the command van topside.

Inside the van, the voices coming through the loudspeaker were loud — but far from clear. The nerve-wracking helium effect was at work on their vocal cords, making accurate comprehension difficult. It was a very serious deep-diving problem that to date, defied solvability.

Bunton strained to make out the words. One phrase he caught several times: "Cold. Very cold." At 7:07, Blackburn reported that Cannon was back in the PTC...complaining of extreme fatigue and coldness. Bunton checked the time. Only 13 minutes in the water and Cannon was done in. Nonetheless, he had completed his primary job which was to open the No. 3 ballast tank valve for flooding, which firmly secured Sealab to the bottom with the added weight of tons of sea water.[9]

Movement flickered on one of the TV monitors. Into view came the shape of a diver, pulling his umbilical behind him. That would be team leader Bob Barth.

[9] On the night of 17 February 1969, when the decision was made at 9:45 P.M. to raise the habitat to the surface, the added tons of water ballast was of grave concern: After the sudden and at that time still unknown cause of Cannon's death only hours earlier, sending divers down (although five were pressurized to 610 feet and available in the other DDC) to make the necessary preparations to blow the excess ballast was adamantly opposed by CDR Jackson M. Tomsky, the on scene commander. The YSD *Marine Boss* (crane barge) would not put more than a 50,000 pound strain on the lifting cable. At 10:30 P.M. with a calculated strain of 45,000 pounds, Sealab III finally broke free of the bottom and was raised to a depth of 70 feet. The YSD would not lift the habitat further until its weight was lessened. During the next three hours (until 1:25 A.M. the morning of 18 February) Bunton along with aquanaut Bill Lukeman, Leading Seaman, RCN, would make a series of four dives preparing ballast tank #3 to be blown-down. Note: On the fourth and final inspection dive, Bunton made a startling, but poignant discovery that for him, at that moment, crystallized the futility — and finality — of Sealab III: He found Berry Cannon's face mask. It was lying on the diving station platform at the foot of the ladder, leading up into the entrance hatch. It obviously had been knocked off during Bob Barth's desperate struggle to lift Cannon to the breathable (18 inch) skirt area gas pocket, or his attempt to force his emergency bypass mouthpiece between Cannon's rigid jaws. Miraculously, the face mask somehow remained undisturbed for 20 plus hours, including the 540 foot lift by the crane. Bunton later gave the mask to Wally Jenkins, Cannon's long time friend and coworker from MDL. [Bunton journal and diving log entries]

There was only one in-water camera at the bottom that gave an exterior view of Sealab, and it wasn't maneuverable. It was fixed on the most critical spot, the entrance hatch area on the underside of the diving station.

The crew in the van watched as Barth settled on the platform suspended just above the ocean floor, about eight feet under the entrance hatch leading into the habitat diving station compartment. He climbed a few feet up the entry ladder, his head disappearing behind the 18-inch-high metal skirt surrounding the hatch he had previously "blown down" (by activating a hand valve allowing compressed gas the same as the mixture within the habitat to remove the water). The skirt area now provided a small pocket that allowed some unrestricted maneuverability. Along with his head, both of his arms were out of the water and for the most part hidden from view. Only the elbows and lower two-thirds of his body were visible.

The figure remained in that position for several minutes. The viewers in the van could see what was happening: Barth was trying to push open the hatch into the habitat, but it wouldn't budge. Although made of steel, it was spring-loaded and should have opened with medium effort — as long as the internal pressure inside the habitat was the same as the water pressure outside.

Only the pressure was not equal. For the last three hours, Bunton had been entering the habitat's differential over-pressure into the log every 15 minutes. The figures had not varied significantly in that time. But at 7:05, as Barth began his struggle with the hatch, "plus 8 PSI" registered on the gauge.

On a note pad Bunton calculated quickly, conservatively using only 6 pounds per square inch DP...the hatch was about 4 feet squared, or 16 square feet...144 square inches to the square foot...that was 16 times 144 times 6...He winced, momentarily stunned at the concluding figure he'd just scribbled: There was more than six tons of weight on top of that hatch!

Barth would never be able to open it; didn't they realize that? Bunton twisted around in his seat and started to say something. But the station commander, a seasoned lieutenant, preoccupied with trying to understand the "quacking" reports coming from the PTC, waved him off.

With puzzled concern, Bunton then turned to aquanaut Lieutenant Matt Eggar sitting next to him and quietly asked if he knew of a new plan...when the gas overcharge was to be shut off? Eggar didn't know, and he appeared as baffled as Bunton. They of course both knew, that it should have been — had to be — synchronized to equalize the pressure on the hatch upon Barth's arrival at the habitat.

Turning back to his screen, Bunton became transfixed by the unfolding events. He saw that Barth was still frozen in the same position. Although the TV monitor didn't show it, he knew that Barth's shoulders and arms had to be trembling from the exertion of pushing upward against the entrance hatch. Damn, Bunton thought, doesn't Barth know it's impossible?

But then, how could he? There was no communications between the diver and the PTC (although commicable full face mask units were available, they were still in a state of disrepair), or the diver and topside. Barth could only assume that the habitat pressure had been equalized. And on that premise, he could only conclude that something else was causing the hatch to stick.

Finally, Barth moved. He had given up. His head came into view again as he sank beneath the hatch skirt, and his hands went to his umbilical. He was hauling himself back to the PTC.

"I almost didn't make it — had a hard time breathing," Barth reported afterward, then adding "that had the PTC been just 20 feet farther, he couldn't have gotten back to it."

But make it he did, at 7:12 p.m., and he immediately went on the emergency BIBS (Built-In-Breathing-System). One hour and 34 minutes later, the transfer capsule was back aboard the *Elk River* and mated to the DDC.

The problem of the helium loss remained. It was now nearly 9 p.m., and the gas supply on both the *Elk River* and the supply barge tied up next to her was getting low. More gas had been ordered but was not expected until seven hours later, around 4 a.m.

September 1968: *After undergoing two years of extensive modifications, Sealab III (now painted yellow) is lowered into the water for the first time at the San Francisco Bay Naval Shipyard. Note the multiple cylindrical configurations on one end, above, and on the diving station/entrance of the habitat, for coiling and outside storage of diver's umbilical hoses.*

(Photo by Bill Bunton)

February 1969: *Off San Clemente Island: Sealab III being prepared for its 610-foot descent commencing at 4:36 p.m. 15 February at a descent rate of 4 feet per minute.*

(USN Photo)

An artist's conception of Sealab III on the sea floor with the Elk River (IX-501) on the surface in support.

(USN Illustration)

An artist's conception of Sealab III on the sea floor with the PTC (Personnel Transfer Capsule) stationed nearby. Note the diving station entrance ladder to the left and Observation Room on the far end to the right.

(USN Illustration)

Late June 1968: *Aquanaut Berry Cannon surfacing after making an open-circuit R&R air dive in the Point Loma (off San Diego) kelp beds.*

(USN Photo)

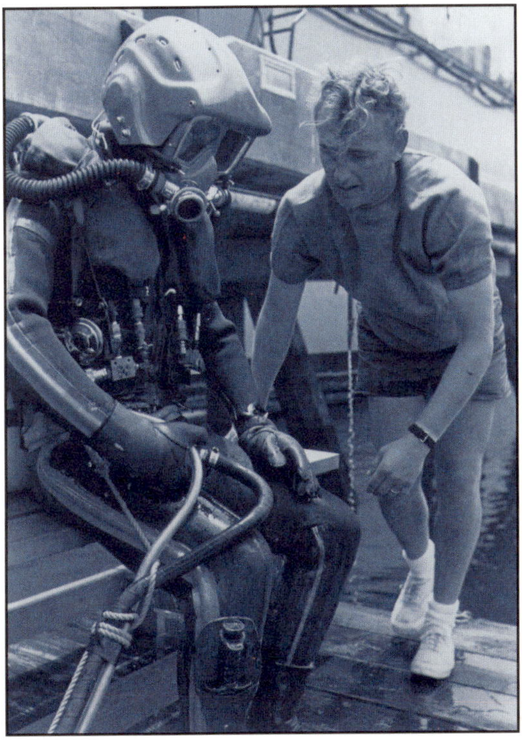

June 1968: *Ballast Point, San Diego: Aquanaut Bill Bunton, being tended by aquanaut Petty Officer First Class Derek "Knobby" Clark, RN, prepares to make a pier-side shallow water training dive. Note the diver is wearing an Open Circuit Hot Water Suit and the MK-VIII Semi-Closed Circuit, Mixed-Gas Breathing Apparatus with a Kirby-Morgan Helmet (total out-of-water weight approximately 135 pounds).*

(USN Photo)

October 1968:
An unidentified aquanaut makes a leaping MK-VIII pier-side training dive at the San Francisco Bay Naval Shipyard.

(Photo by Bill Bunton)

November 1968: *San Francisco Bay Naval Shipyard: Left to right, aquanauts Berry Cannon (Team I) and Wally Jenkins (Team IV). Note the Sealab III Observation Room viewing port in the background.*

(Photo by Bill Bunton)

September 1968: *Sealab III entering the water at the San Francisco Bay Naval shipyard for the first time.*

(Photo by Bill Bunton)

February 1969: *An aerial view of the Sealab III support vessel Elk River (IX-501), under way for San Clemente Island.*

(USN Photo)

Above, an artist's conception of two Sealab III aquanauts exiting the 7-foot diameter PTC (Personnel Transfer Capsule).

(USN Illustration)

February 1969: *Off San Clemente Island: One of the two HY-80 steel PTCs (Personnel Transfer Capsules) being lowered into the water through the center well of the Elk River for a check-out dive.*

(USN Photo)

November 1968: *Phillipe Cousteau (from France), one of five more recently selected aquanauts who will participate in Sealab III, and assigned as the ninth member of Team III, inspects the habitat. Below, Cousteau being tended prior to making a training dive at the San Francisco Bay Naval Shipyard.*

(Photos by Bill Bunton)

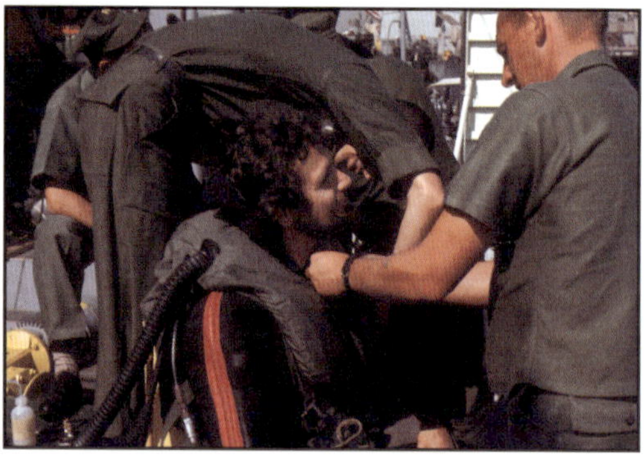

8 February 1969: *San Clemente Island (North Point): Returning from a training dive are (left to right) aquanauts LT Laurence Bussey, USN, the scheduled Team 1 photographer, and Wally Jenkins, who is learning to use Bunton's 400-exposure camera.*

(Photo by Bill Bunton)

Sealab III, Team I Aquanauts: *From left to right kneeling: Berry Cannon, civilian; Jay Myers, Machinist's Mate Second Class, USN; LCDR James Vorosmarti Jr., MC, USN; Richard Cooper, civilian. From left to right standing: George Dowling, civilian; Richard Blackburn, Aviation Ordnanceman First Class, USN; Warrant Officer Robert Barth, USN, Team Leader; John Reaves, Photographer's First Mate Class, USN (replacement for Lt. Laurence Bussey); and Richard Bird, Engineman First Class, USN.*

(USN Photo)

May 1979: *George Foote Bond, Captain (MC), USN, retired — as he appeared three plus years before his death on 3 January 1983.*

(USN Photo)

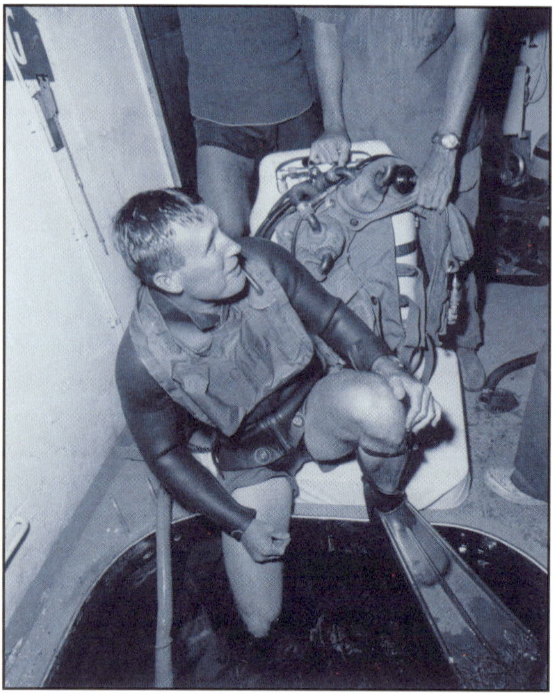

September 1968: *Aquanaut Berry Cannon, sitting in the diving station entrance hatch of Sealab III, prepares to make a shallow water MK-VIII training dive at San Francisco Bay Naval Shipyard.*

(USN Photo)

18 February 1969: *During the early morning hours off San Clemente Island: After a second aborted dive to gain entry to the habitat and the death of aquanaut Berry Cannon, Sealab III is finally lifted to the surface.*

(Photo by Bill Bunton)

Sealab command had a decision to make. The start of Sealab III had already been delayed several times. The Navy had finally committed itself to a certain date, 17 February. It was now the night of the 16th. At least a dozen reporters waited on the island to cover the event. Another delay would once again, be terrible publicity that at all cost — had to be avoided.

The choice was to go down again as soon as possible and give it another try.

Within an hour, at approximately 10 p.m., Sealab command made another choice: The same four-man team that had just come back up would go down again, despite the fact that a second team, compressed to the 610-foot bottom depth at the planned rate of compression since 10 a.m. that day, rested and fresh, was still on standby in the other DDC.

Both decisions tempted fate beyond endurance.

17 FEBRUARY THROUGH 13 MARCH 1969
SAN CLEMENTE ISLAND AND SAN DIEGO

It was now 2 a.m. The Robert Barth/Berry Cannon-Richard Blackburn/John Reaves team (all but Blackburn were veterans of Sealab II) had spent the past five hours in the warmth of the DDC, trying to recover from their nearly four hours in the numbing cold environment of the first dive. Barth and Cannon were additionally exhausted from heat loss and physical exertion. However, this was the more experienced pair, having trained for weeks in the entry procedure. The choice was made sometime after midnight that they, once again — and not Blackburn and Reaves — would have the better chance of gaining entry and quickly solving the leakage problems of the Sealab III habitat on the ocean floor.

This would turn out to be a *fateful* decision.

Barth didn't seem happy about going back down. He had complained of pain and difficulty in breathing during the first dive. Cannon didn't say much about what he'd just experienced, but then Cannon seldom did if it went against command authority. He was a knowledgeable, seasoned diver with an indefatigable work ethic — and possessed an unshakable faith in the Navy's ability to make the right decisions.

When Bill Bunton was relieved of his duty station at 11:50 p.m. (16 February) and back in his quarters on San Clemente Island 90 minutes later, the "choice" of which two of the four aquanauts from the first dive would be asked to make the second entry attempt, had not as yet been determined. But from topside command's perplexing indecisiveness he'd just witnessed earlier in the night, Bunton was mentally drained; and with each passing, sleepless hour, more apprehensive than ever....

During Sealab II, the divers had been able to use Mark VI semi-closed-circuit scuba units that allowed the aquanauts the freedom of not being tethered to their breathing source. These units were not feasible at deeper depths because the tanks of helium-oxygen mixture carried on the divers' backs simply could not hold enough gas to last any reasonable length of time. Although duel tanks (each containing 90 cubic feet at 3000 psi) would be worn on all excursion dives from the habitat, their primary use would be under emergency conditions only.

The Sealab III solution for breathing-gas conservation was to use umbilical hoses (200 feet in length) between the diver and the personnel transfer capsule (PTC), and between the diver and the habitat with an umbilical option of either 200 or 600 feet. It was (theoretically) a workable system, in principle very much like that used in space walks, only heavier. But as the Navy was to discover only too soon, there was a significant difference between an umbilical suspended in weightless space than one dragging across unknown obstacles on the ocean floor.

It was now 2:50 a.m. The team of Barth, Cannon, Reaves and Blackburn already had received their orders for the second dive.[10] During the interim they'd had hot soup, some rest, but little sleep — an hour at the most. As they prepared for their

[10] Commencing at 2:50 a.m. through 9 a.m. 17 February 1969, all times, events, numbers, quotes and/or paraphrased comments are either (1) transcribed from the official log (Form MDL 8500./1 [8-68]) that was kept in the command van "communication

Continued Next Page

second descent to Sealab III, the flood gauge in the habitat sounded the alarm: Water had reached a depth of six inches, due to a slight negative differential pressure (DP) presently being maintained within the habit.

At 3 a.m. the new gas barge arrived. The supply was enough for only 18 more hours at the present rate of loss. Nevertheless, Sealab command went ahead with its plan.

At 3:40 a.m., the PTC carrying the diving team detached from the DDC. At 4:13 it entered the water, and at 4:33 it reached bottom depth. Thirty-five minutes later, Barth and Cannon opened the hatch and swam out. At 5:08 a.m., Barth's figure appeared on the TV monitor up in the van. Cannon could not be seen; presumably, he was working at the other end of the habitat, out of sight.

The TV camera's flood lamps illuminated scores of fish against the black background of the water as Barth climbed the ladder to the diving station entrance. Again, he tried to push open the hatch that had been an impossibility on the previous dive.

To the men topside, the picture on the TV screen was, at first, almost a replay of what had happened on the entry attempt just a few hours earlier: Barth strained in vain, gave up and swam out of view. Then it became a riveting drama they were watching on the monitor.

A squeaking voice from the PTC says Barth has returned to pick up a crowbar. The PTC is behind the habitat, hidden from the camera. The camera shows only the forward underbelly of the habitat, with the ladder extending down from the entrance hatch, and the sea bottom for five or ten feet beyond. The water has turned murky because Barth's trailing umbilical has stirred up the bottom sediment.

Barth reappears on the screen, the crowbar visible in his right hand. Suddenly he switches direction, swimming rapidly away from the camera, and out of view to the right. His flippers stir up more clouds of silt.

---

watch"; (2) transcribed from personal notes and/or conversations with other on-scene aquanauts; (3) subsequent direct testimony witnessed during the 28 February - 12 March 1969 Board of Inquiry hearing; and (4) viewing of the second dive video film during said hearings. [Bunton journal entries]

"Holy shit..." The strangled words come from one of the men in the command van. The others are silent, staring mesmerized at the screen.

Barth reappears. He is towing something.

"It's Berry," someone whispers.

Barth is seen dragging Cannon to the diving station's support platform. He props him upright against the ladder, apparently trying but failing to get his head within the skirt's breathable gas pocket. Barth then attempts to give Cannon gas from Cannon's demand-type emergency Aqualung regulator (attached to all Mark IX units). More clouds rise as Barth works, further obscuring the vision topside. Regardless, those watching helplessly, know what is going on. Barth is trying to force the mouthpiece between Cannon's teeth.

"Get another diver in the water now — and send him to the habitat!" The diving officer in the command van snaps the order down to Blackburn and Reaves in the PTC. Acknowledgment crackles over the speakers.

Bob Barth continues to struggle, desperately. Cannon is now convulsing. Within seconds that seem endless to those topside, the team leader finally gives up and swims off, dragging Cannon with him. They both disappear from view.

Then quite startlingly, Barth is back again on the screen — alone. As he reaches the ladder it becomes obvious why he returned: His umbilical is fouled in the ladder rungs. He doesn't have enough free hose to make it back to the PTC. He labors to untangle it, his motions revealing that he is all but physically spent. Finally, he dislodges the umbilical and with short, exhausted strokes, swims off in the direction of the PTC.

Berry Cannon cannot be seen in the peripheral darkness, even after the silt settles behind Barth. The aquanauts watching the monitor have no way of knowing that Cannon is lying only a few feet from the PTC. Barth, after freeing his umbilical snarl, doesn't have enough strength left to pick Cannon up.

As Barth drags himself to the PTC, Blackburn, following the orders from topside, is just getting out. Barth is partially hauled inside by the remaining diver, Reaves. Within seconds,

Blackburn locates Cannon and is at his side. Blackburn, six feet four inches and built like a bear, assisted as much as possible by Barth and Reaves, picks up Cannon with both hands and thrusts him up through the open hatch of the PTC.

It is 5:15 a.m. Only seven minutes have elapsed since Barth first appeared on the TV monitor. How long Cannon has been in trouble, no one knows. It could have been the full seven minutes, or it could've been only three.

It doesn't matter. At 610 feet, even one minute is too long.

The PTC reports by phone that Cannon "is in bad trouble." Reaves and Blackburn try to revive him, first by forcing gas (utilizing the BIBS) into his lungs and then by cardiac massage and mouth-to-mouth resuscitation. They continue their efforts as the PTC is being hauled up as fast as the gantry crane winch can handle the load. During the ascent, Cannon does not regain consciousness. He is not breathing. His three comrades think he is dead. Though they are not certain during the horror of the moment.

At 6:05 a.m., the PTC is hauled back aboard the *Elk River* and mated to the DDC. Cannon is placed in the outer air lock. If he is still technically alive, he's brain-dead and has been for nearly an hour.

No doctor on the outside can quickly enter the pressurized DDC (although LCDR Mark Bradley MC volunteered) without someone inside first closing the hatch that separates the two chamber locks (compartments), then venting the outer lock, then once again pressurizing it to the 610-foot depth.

A time-consuming procedure that would not even have existed — *if* the smaller outer lock had been left open to normal atmospheric pressure prior to commencing the dive, as is the standard practice. An emergency entry then could have been made in less than half the time, or about ten minutes.

But then again, "if" the outer lock had been available, medical assistance had more than enough time (at least 45 minutes) to gain access to the inner lock and awaiting the arrival of the ascending divers — with *no loss* of time. However, this impossibility to close the hatch separating the two vacated chambers, was not the reality of the situation.

If indeed time was crucial during this chaotic hour or so since topside command was first alerted to the escalating emergency taking place below, the practical solution was only a few feet away. Doctor/aquanaut LCDR James Vorosmarti MC, in the adjacent DDC and already pressurized to 610 feet, could have readily been transferred via his own PTC, prior to the arrival of Barth's team. *None* of these options were exercised.

Either way, apparently, it is far too late. After conferring with team leader Barth and from closed-circuit television and porthole observations only, all concerned topside authority are convinced Berry Cannon is dead.

Cannon's body, still pressurized to bottom depth, could not quickly be brought to atmospheric pressure without triggering "explosive decompression." (A condition of aeroembolism characterized be the sudden and rapid formation of inert gas bubbles [helium in this case] within the blood, manifesting in morbid distention of the lungs, organs, and skin of the body.)

Notwithstanding the post-mortem results, a few hours later the difficult, though expedient decision is made: Commencing at 9 a.m. the outer lock is vented in less than 60 minutes, and Cannon is returned to the surface.

Searching for Answers

Bunton couldn't believe what he read in the newspapers the next day. The *Long Beach Independent* for 18 February 1969, under the headline "Diver's Death Halts Sealab," reported:

"San Clemente Island — The nation's most ambitious Man-in-the-Sea experiment, Sealab III, was postponed indefinitely Monday after one of the Navy's best aquanauts died of an apparent heart attack in bone-chilling waters 610 feet below the surface.

"In a shoreside interview, project manager Captain William M. Nicholson, tired and disappointed, announced suspension of the planned 60-day undersea venture.

"Dead is 33-year-old Berry L. Cannon, a civilian [working for the Navy] electronics engineer who was the father of three boys.

"Nicholson said the detection of several helium gas leaks in the 63-foot-long Sealab habitat was another factor affecting the decision to delay the project.

"The [surviving] aquanauts in the two deck decompression chambers (DDCs) on the surface support vessel, the *Elk River*, are undergoing decompression—a process that will require them to remain in the chambers for approximately seven days.

"Nicholson said the habitat was being readied to be brought back up to the surface, and 'We will not resume diving until we learn what caused Cannon's death.'

> "He said Cannon's body was removed from the DDC through a lockout chamber and taken to San Diego Naval Hospital. 'We do not know whether Cannon died as a result of making the dive or whether it was a coincidence that he suffered a heart attack at the very moment he was making the dive,' the captain said."

That made it all but official. Cause of death: heart attack. Other Navy personnel began to use the phrase themselves, as though it were fact. Two days after the tragedy, however, a naval Man-in-the-Sea spokesman told the press, "We have been informed that the gross autopsy was not conclusive. Laboratory tests are continuing."

San Diego medical writer Cliff Smith was knowledgeable enough to interpret the statement. "This," Smith wrote in the *San Diego Union*, "seemed to rule out the possibility that Cannon died of coronary occlusion, the most common cause of sudden death and what is usually meant by 'heart attack.'"[11]

But by this time, the Sealab story had moved off the front pages of newspapers across the country. What most people remembered was the first-day headlines that in an unfortunate coincidence, a diver had suffered a heart attack while he happened to be on the ocean floor.

Sealab III remained suspended pending the official Board of Inquiry hearings. With proceedings closed to the press, the board convened in San Diego (at the Point Loma based Naval Undersea Warfare Center) on 28 February, and concluded on 12 March. There, the startling truth came out: Berry Cannon *had not* had a heart attack. He died of carbon dioxide poisoning.

Someone — inexplicably — had neglected to fill the carbon dioxide scrubber on one of the four Mark IX diving rigs previously placed in the PTC. Cannon, after his first few breaths upon entering the water, had been breathing pure poison — his own carbon dioxide–charged exhalations.

The scrubber was an aluminum canister hooked into the semi-closed-circuit breathing system. Before each dive it was filled with a fresh load of eight pounds of baralyme granules.

---

11 Smith, Cliff. "Cause of Aquanaut's Death Eludes Probers," *San Diego Union*, 19 February 1969, sections A-1 and A-4.

Each exhaled breath would pass through the canister, be cleansed of carbon dioxide by the baralyme and pass into a breathing bag. This bag in turn provided the diver with his next breath, supplemented by a small charge of fresh gas from the umbilical attached to the diving unit and connected to the PTC source.

Cannon's canister proved to be completely empty. It had *never* been filled.

The four Mark IX diving units, each one weighing 30 pounds were supposedly set up, checked, and rechecked by other aquanauts, had been placed inside the circular wall of the PTC. Upon entering the capsule, each diver grabbed the nearest unit.

One of the four divers that morning was going to die. That Berry Cannon was the one who'd taken the unit with the empty canister had been pure chance.

On 19 February a plane was chartered for 12 aquanauts, including Bunton, to attend a memorial service in Chula Vista, California for their fallen comrade.

It should not have happened. Theoretically, it could not happen. There were too many safety checks in the procedure for setting up the rigs. But happen it did.

Who was to blame? Medical Officer Bond's anger and frustration was evident when he was quoted in the media as saying: "The project had been strangely trouble-ridden for more than two months before the fatal incident." He also told the press: "...to my knowledge, never before has our program had anything like the sheer number of equipment failures seen in Sealab III. Malicious or otherwise, the gas leaks and other hardware mishaps were, to me, just unbelievable."[12]

Bunton, like the rest of the once-elite corps of aquanauts, returned to his original work facility at NUWC, where he'd been promoted (in absentia) to Diving Supervisor since May of 1967.[13]

---

12 Stocking, Susan. "Sealab's Difficulties Called Unbelievable," *Los Angeles Times*, 6 March 1969, Part I, 3 and 32.

13 While in Washington D.C. on 3 May 1967 at the Navy Experimental Diving Unit (EDU) to make the first of two saturation dives (200 foot and 600 foot), Bunton was informed in person by George Anderson (Asst. Dept. Hd. Ocean Sciences Dept.

Continued Next Page

Yet as much as Bunton now wanted to erase from his mind the ill-fated disaster and previous three years, he couldn't seem to forget captain Bond's recently quoted statements to the news media. What could he (Bond) possibly be implying by using the words "...strangely trouble ridden..." and even more disturbing "...*malicious* or otherwise...were to me just unbelievable..."? (Bunton would wait a long time to find out.)

The Board of Inquiry sat for almost two weeks yet was unable to pin down the responsibility. The official finding was death by carbon dioxide poisoning. No definitive blame settled on anyone.

The day after the inquiry closed, Wally Jenkins phoned Bunton's San Diego office from Panama City. They'd gone through Sealab II together and, had fate not entered the picture, would have been diving buddies again on Team Four in Sealab III. Together almost continuously for the past four years, and both being civilians in the Navy-run project, they'd become fast friends. Jenkins wanted to know what had taken place at the hearings.

Bunton sidestepped: "Didn't the newspapers there cover it?" He was uncomfortable. Jenkins had been close to Cannon, too, and as an oceanographer, Jenkins had worked with

NEL), that, for his contributions to (1) the success of Sealab II and (2), his continuing efforts to enhance and expand diving support capabilities in behalf of NEL, he was being promoted to Diving Supervisor. This appointment was formalized in MEMORANDUM 3102-733-67 by Dr. G.H. Curl, Coordinator of Ocean Sciences NEL, on 31 May 1967. Note: Although a portrait (plaque) of Bunton would soon be hung in the lobby of "Building 33": (San Diego Administrative Headquarters) as only the third recipient of the "Navy Superior Civilian Service Award" ever presented to a NEL employee through 1965, his ongoing ambitious plans to modernize the laboratory's diving capabilities had become an inner-department threat to certain egos. And in less than two years, with his mentor and always encouraging supporter Dr. Curl now dead only a few months, Bunton quickly found out that he too, would become a casualty of Sealab III within days of its collapse. On 3 March 1969 at 3:45 P.M., after the Board of Inquiry adjourned for the fourth day, Bunton's supervisor Art Schlosser, Hd. Ocean Eng. Div. of the Ocean Sciences Dept., summoned him to his office: At that time Bunton was informed he no longer would be the (highly coveted) Diving Supervisor (of now renamed NUWC). Schlosser's explanation was that Bunton had ...although on official government orders assigned (temporarily) to the Sealab project... been gone or absent from his job and principal responsibilities just — too long. Stunned, humiliated, and bitter beyond belief by Schlosser's ignominious rationalization for his demotion, upon completion of his attendance obligation to the still ongoing Board of Inquiry hearings, Bunton resigned his civil service position and left the Naval Undersea Warfare Center on 3 April 1969. [Bunton journal entry]

Cannon for many years at the Navy Mine Defense Laboratory in Panama City. Jenkins hadn't attended the hearings because he'd been ordered back to Florida within hours of Cannon's death.

"Sure they did, front page every day," Jenkins said. "Though that's not the same as being there. You went, didn't you?"

"Yeah, I had to. I was supposed to testify. They even told me to prepare a written statement.[14] But they never called me to the stand...I don't think they liked what I wrote."[15]

"The papers said the only disciplinary action was a letter of reprimand to (designated Commanding Officer Jack) Tomsky," Jenkins said. "I know he had to get one; he's the honcho of the whole affair and has the responsibility for whatever happens. Everyone knows it doesn't mean a thing."

"True," Bunton agreed. "And we both know there were a dozen other aquanauts and boot-camp Seabees, recently recruited only some hours earlier, in and out of the diving locker area.

---

14 When Bunton returned to SCI (San Clemente Island) 20 February 1969, after attending a memorial service for Berry Cannon in Chula Vista, Calif. the previous day, and arranging a bank "Trust Fund" account for the benefit of Cannon's widow Mary Lou, he was informed he had to prepare a written statement of the facts as he remembered them, leading up to and surrounding the death of Cannon. Also, he was advised, he would personally be called to testify during the Board of Inquiry hearing that was to convene in San Diego on 28 February. Again back in Long Beach, on 24 February after having coffee with Scott Carpenter, he later turned in one of his two identical seven page hand written copies of his statement at the Long Beach Airport headquarters of CIB (Sealab's Command Information Bureau). Note: (1) Bunton gave this enveloped statement to "Greenwood" to be given to Yeoman Billy Snell for typing. (Both were USN personnel working for and attached to CIB); (2), attached to his then — "draft only" statement — Bunton specified that as written, it was still confidential and once typed, it was to be returned to him for his final edit, before being submitted to Sealab III command or the Board of Inquiry; and (3), after waiting four days for his statement to be typed, on 28 March, only hours before the Board of Inquiry was scheduled to convene, Bunton demanded his original hand written draft be returned to him. It never was. Exactly where it finally wound up is not known. However, he still retains his second original which, due to the lack of readily available "after hours" copying facilities (in 1969), was quickly written the night of 23 February at the Long Beach "Outrigger Motel." [Bunton journal entry]

15 The high ranking naval Board of Inquiry briefly convened 28 February 1969 (closed to the public and media) at the Naval Undersea Warfare Center, San Diego, and officially, commencing the next day 1 March (Saturday 6 P.M.), held consecutive daily hearings including testimony, through 12 March. Bunton attended all sessions, and was never (as previously told he would be) called to the stand to testify. [Bunton journal entry]

"The Navy had given the press a final deadline date ... that Sealab III would commence 17 February, so that's when we had to start. No more delays, no matter what the cost or how inexperienced the help. You saw those 13 sailors they dumped on us from the Naval Training Center — not one ever even had a mouthpiece between his teeth, and they were running all over the place! Christ, they were only kids trying to help out, flown in at the last minute when everything was becoming desperate...

"The Navy couldn't blame any of the Seabees. It wasn't their fault. Everybody just had too many damn things to do at one time, especially in those last 36 hours."

"But didn't that come out at the hearings?" asked Jenkins.

Bunton's voice dropped. "There were a lot of things that didn't come out ... or were not thoroughly investigated."

Jenkins' voice dropped, too. "Like what?"

"Like the fact that they roped off the DDCs the day after Berry died and put guards around them."

"Jesus, why on earth would they do that?" Jenkins wondered.

"Because they found an open oxygen valve on the line leading up to the emergency Built-In-Breathing-System [BIBS] in one of the chambers — while the guys were still in there decompressing."

"An open oxygen line?" Jenkins sounded incredulous. "One day after the dive? C'mon, Bill. They must've only been at the equivalent of 400 feet into their ascent. Pure oxygen at that depth would've had them convulsing in three minutes, dead in ten!" (Breathing 100% oxygen at depths greater than 33 feet becomes increasingly toxic.)

"They were lucky. Only the front valve to the main gas supply was open. The other one down the line was closed. And you won't believe this: It happened twice."

"You mean someone *deliberately* tampered with it?" asked Jenkins.

"Even worse. After the first time, that oxygen valve was wired closed. So for whatever reason, someone broke the wire seal and opened the valve again. There was such concern that a 24-hour guard was stationed around both DDCs, and an officer had to ascertain the valve lineup every hour during the remaining days of decompression."[16]

Jenkins wanted to know more about the testimony, and Bunton sketched some of the highlights: The listing of the numerous mechanical problems, the screening of the videotape showing Barth's fouled umbilical — and the lack of adequate deep-water aquanaut training, the project's most serious failure from the beginning.[17] "Despite two weeks of hearings,"

---

16 (1) As conveyed 25 February 1969 to Bunton over dinner by aquanaut Richard Cooper, whom he'd just picked up at the Long Beach Airport (arriving from San Clemente Island); (2) again with Cooper on 26 February and both were joined by George Dowling (also a member of Team I who'd just completed his seven days of decompression aboard the *Elk River* on 24 February, 1:30 P.M., and was now in Long Beach en route to his home in Panama City, Florida) who further elaborated on this — "inconceivable happening"; and (3) as only partly witnessed (on scene) by Bunton who left SCI 5: P.M., 18 February to attend Berry Cannon's memorial service in Chula Vista, Calif. and did not return until 11:30 a.m., 20 February, then left SCI once again (for the last time) for Long Beach 9 A.M., 21 February. [Bunton journal entries]

17 Board of Inquiry testimony 1969, as stated by three Team I aquanauts who made and survived the first and second — and only — 610 foot dives to the Sealab III site: (1) 2 March, John Reaves, USN ... "previous training dives using the MK-IX (semi-closed circuit mix-gas breathing apparatus) only made at depths of 30 to 50 feet." (2) 2 and 10 March, Richard Blackburn, USN... "MK-IX not breathing right...wanted to use MK-VIII (semi-closed circuit mix-gas breathing apparatus) instead...poor training." (3) 2 and 10 March, Robert Barth, USN... "Problems all along with Sealab III...not much confidence in equipment...hard time breathing...constantly on MK-IX by-pass." (4) Regarding Sealab III's aquanaut (required for most but not all) training program, and based on Bunton's participation: (a) Attended (included among a class of 40 other civilians, Navy officers and enlisted personnel) the U.S. Navy School for Deepsea Divers in Washington D.C., 10 October 1966 through 16 December 1966. During ten weeks of training, all dives were made inside the school in either shallow water wet-pots or pressurized wet chambers, except three dives in the adjacent Anacostia River at a maximum depth of 30 feet. (b) In June 1968 (17-months later) aquanaut diving training again commenced in San Diego Bay (Ballast Point), for the first time utilizing the MK-VIII. Bunton (during June) made his first check-out dive in a swimming pool, then 12 more, all pier-side, at depths of 40 to 50 feet. (c) Two more MK-VIII training dives were made 2 and 3 September 1968 at the San Francisco Naval Shipyard. Both dives were also made pier-side, at a maximum depth of 40 feet. (d) At the same location (and again pier-side), commencing 27 September 1968 through 8 October, Bunton made three dives for the first time utilizing the MK-IX, at depths of 40 and 50 feet. (d) At San Clemente Island commencing 29 November 1968 through 29 January 1969 (from the IX-501 *Elk River*) concluding his training with the MK-IX, Bunton would make four more dives. These were brief, five minute lock-out dives from one of the two PTC's (Personnel Transfer Capsules), at depths of only 20 to 35 feet — and his first made in the open sea during all Sealab III training. (e) On 5 February 1969 Bunton would make one more open sea dive, utilizing the MK-VIII, to his deepest training depth obtained — 70 feet. Note: By contrast to the sporadic Sealab III shallow-water, "pier-side" (basic) training detailed herein over a two-year time span, all 28 aquanauts selected for Sealab II

Continued Next Page

Bunton said, "they never did figure out who had neglected to refill the empty baralyme canister."

"I guess it really doesn't matter," Jenkins said slowly. "What matters is the way they set up the system. Berry had no control over his own equipment. He had to rely on someone else."

In Sealab II, each diver had usually set up his own diving rig. But in Sealab III, because of the usage of DDCs and PTCs, and because of the topside manpower shortage that forced continual rushing and little sleep the final 36 hours, rigging responsibilities went to just about any other aquanaut except the diver involved.

Promising to stay in touch, Bunton hung up. Then he rummaged through his desk for a report he'd given on the "Problems of Deep Underwater Photography." Captain Bond had asked him to present the topic at a Washington D.C. symposium on "Man's Extension into the Sea" in January 1966, after the completion of Sealab II. His oral presentation was later printed in the symposium volume of technical papers.

Bunton scanned his article until he found the paragraph that was banging around in his head:

> "...Saturation dives from a pressurized habitat ... can be dangerous because of the complexity of the self-contained units used for mixed gas. Unfortunately, the problems that can develop with this equipment are such that there is little warning, if any, and a diver must constantly be on guard for the slightest sign of a malfunction. In many instances, others may have set up his unit for him; this can be a disturbing thought...On a saturation dive, he now knows that in case of an emergency he cannot surface without meeting certain death. He can no longer take his diving apparatus for granted, and years of training and his own instincts are contradicted. This condition creates a mental hazard that burdens the diver's mind and, in varying degrees, can detract from his functional capabilities..."

---

were intensively and realistically trained in the usage of the MK-VI semi-closed circuit mix-gas diving apparatus (to be their primary diving unit during Sealab II) miles offshore, in the Gulf of Mexico. Assembled at MDL (Mine Defense Laboratory) in Panama City, Florida 5 May 1965 through 27 May, over the next 17 days Bunton would make a series of 19 progressively deeper dives, culminating with the last three made to 180 feet. [Bunton journal entries, diving log, and copy of his hand-written, seven-page statement prepared for the Board of Inquiry]

Bunton quietly closed the volume and put it away.

But once again, he could not avoid thinking what now must be convincingly evident to all: That the professional planning for Sea Lab III just wasn't there. If Berry Cannon hadn't died on that dive and the way things were progressing, it likely could've been some one else on the next, or myself on the Fourth Team. Or, conceivably — a whole team!

Although Bunton tried during the last few months to speak out as to what he foresaw as evolving problems, both verbally and in writing, his voice as a civilian only, had no value in the hierarchy of naval decision makers.[18]

With one last irrepressible "if only" thought of prophetic hindsight he could rewrite his symposium paper of three years ago — despite the certainty it would end his continuing participation in the Man-in-the-Sea program — the conclusion would read: At this still dangerous experimental state of diving to such extreme depths, the primary skill of all of us, as an aquanaut or topside control, must unquestionably be as highly experienced and dedicated divers first. Other skills, rank or those with ambitions better suited for shallow-water endeavors, should be of secondary importance — not the other way around....

APRIL AND MAY 1981...

No longer restrained to fully express his inner most thoughts, Captain Bond (now retired from the Navy) wrote Bunton three separate letters within a four week period.

The once proud chief of Sealab I and II would never forget nor forgive those he considered responsible for the dire circumstances that lead to that inglorious night, that ended Berry Cannon's life.

18 As conveyed or discussed with various "positioned" civilian personnel associated with Sealab III, and Navy officers (and non-officers) directly or indirectly connected to the project, including but not limited to the Washington D.C. "certification team," CAPT George Bond, CAPT Walter Mazzone, CDR Carpenter, LCDR Mark Bradley, LT Laurence Bussey, and LT Matthew Eggar. [Bunton journal entries and remembrances]

The following are stunning excerpts in Bond's own words (italics and brackets provided by the author) that finally clarify his long ago ambiguous statements to the news media:

> ...Three important facts have never been made public to this day — and you had no way of knowing them... I will brief you numerically:
>
> 1. After innumerable material casualties [here Bond refers to an associated on scene Sealab III naval officer], and I quietly decided that one person was out to delay or *actually sabotage* the program. The two of us went through a long list of all participants, scientists, contractor employees, and actual program personnel. All factors pointed to one person — a contract engineer who on several occasions in drunken moments expressed his dislike for this program which trapped him on that desert isle [San Clemente] and to which he had been assigned against his will. After each equipment failure, his reaction was always adversary in nature; most importantly, he was always on the scene when the accidents occurred...
>
>   In all events, the system of each aquanaut filling his own carbon dioxide absorbent canister went by the board when Team One was ordered to depths post-haste. On this day, canister filling could have been done by Seabees, but our suspect contractor employee was seen in the mixed-gas preparation cage where he had no business.
>
>   Finally, after the tragedy, this character reportedly made remarks which were grossly out of order. Of course, *not one word* of this was allowed in testimony at the [Board of] Inquiry.
>
> 2. I don't believe you knew it at the time [Bunton did know this], but the man who controlled the internal pressure of the habitat was in your same command van. When Barth failed on his first attempt to open the hatch, he eased pressure at first, then saw water coming in the diving station where it could do no harm, and reacted by increasing the internal pressure back to (six) psi differential. I saw what happened from our upper-deck video scan, and at our conference after the first attempt, I asked permission to tell that operator to drop internal pressure at the time appropriate for opening of the lower hatch. This was not granted, but I was informed it would be taken care of through the command chain. This, of course, *was not done, and the mistake was repeated again on the fatal dive.*
>
>   Just two other points: One week before the dive, Washington suddenly denied us use of electric underwear. We jury-rigged the hot water suits, and you saw the results [heated water turned as cold as the sea before reaching the suits]. Next, the electric heaters in the PTC were deactivated without my knowledge, and the duct baffles [for circulating the warmed incoming breathing gas] were removed by *persons unknown* and no record made of this action.

> 3. Finally — and this to me was the worst aspect of the whole mess — we, the parties of interest, were told by the Captain who ran the Inquiry [into the fatality] that under no circumstances would he permit testimony relative to anything not immediately and directly connected with Berry Cannon's death. Thus, we could not even mention the lousy quality control which caused the leaks, the skipped test procedures which brought us to those last days in an unready state, or the fact that gas leaks in the main umbilical (from the surface) *guaranteed a power failure* in the habitat within a matter of hours. Apparently, such testimony would ruin careers all the way up to very senior officers in Washington, who had made *rotten decisions from day one*, and asked for and got a cover-up through distorted military jurisprudence...

Inasmuch as one other Sealab officer apparently concurred with Bond's suspicion of "sabotage" and had not acted on it or come forward with such a heinous charge, Bunton concluded: They, in their combined disbelief of events throughout the night of 16 February 1969, still could not rationally accept that it was the inevitable result of — abject blundering that realistically caused Cannon's death and the collapse of the project.[19]

Nonetheless, however troubling Bond's letters remained, Bill Bunton would never learn exactly who Bond suspected of such malevolent behavior during the final days of Sealab III; or, if in fact his allegation was true.

On 4 January 1983, the *San Diego Tribune* published a highlighted obituary that quickly became known throughout the worldwide diving community:

"SEALAB'S PAPA TOPSIDE IS DEAD"

The obituary continued...

"Captain George F. Bond, who pioneered the 'Man-in-the-Sea' concept, the theory that humans could live on the bottom of the ocean, died yesterday in a Charlotte, North Carolina hospital. He was 67 years old. His military service spanned a total of 33 years...."

---

19 It is the considered opinion of the author that CAPT Bond's written words should speak for him alone...And the concurrence of his quoted thoughts (in part or whole), by the revealing of said officer's name, is of no relevance herein. [Author — Bill Bunton]

On 17 May 1991, the world's newest, ultra modern hyperbaric "Ocean Simulation Facility" located at the Naval Coastal Systems Laboratory in Panama City, Florida honored him by emblazoning his name above its entrance.

Bond's personal journal was published posthumously in 1993 by the Naval Institute Press (NIP) under the title: *Papa Topside — The Chronicles of Capt. George F. Bond, USN.* Edited by Helen A. Siiteri.

Even though Bond was well known to be a prolific writer, covering all significant events of the Sealab years — daily — strangely absent from the NIP publication is his aforementioned, astounding thoughts and revelations.

This was (is) especially disturbing, when Bond's last letter to Bunton of 21 May 1981 (written from Tega Bay North Carolina) concluded with:

> "*All of this* will be in my book, documented as best I can. Meanwhile, as I have said previously, if you can use what I have just written, feel free to do so...or to use my name anytime, any place...I have complete trust in your judgment."

Was Captain Bond's meditative writings considered just too controversial for public consumption — and abridged?

One now can only speculate that if Papa Topside had lived to oversee the publishing of his own journal, would the historical record (truth) of the U.S. Navy's Sealab project be better served — or left as is?

SIXTEEN YEARS AFTER THE SHATTERED ILLUSION OF SEA FLOOR HABITATION AT GREAT DEPTHS...

> "The entire Sealab project," wrote Bill Barada in the September 1985 *Skin Diver Magazine*, "was constantly hampered and harassed by opposition from indifferent and incompetent Navy officials ...[After] the Navy bureaucracy took charge, decisions and procedures were dictated by Washington brass, and experienced Sealab personnel were ignored. The result was a long series of failures that ended in disaster.
>
> "Cannon's death was an excuse to abort the entire program. The project was mothballed, equipment and materials were stripped, and

nothing remained of the $10 million Man-in-the-Sea program except a rusting steel hull.

"The high hopes and expectations of U.S. Navy aquanauts for a successful conquest of inner space went down the drain."

THIRTY YEARS AFTER THE DEATH OF AN AQUANAUT...

As critically accurate as Barada's long ago summation may be, it also should be appreciated and remembered that this was a human endeavor...an epic attempt in the most hostile environment known to man. Certainly plenty of serious mistakes were made, through inexperience, naiveté, or other reasons of speculation...

But, as in all death-defying experimental projects, lives and careers brimming with vigorous dreams of lofty achievement, have been lost out there on the edge. The aquanauts all knew this, yet were willing to accept the dangers of the deep.

To taste your own fear and apprehend the high risks...and still push past the edge...is exactly what Berry Cannon did during those fateful early morning hours of 17 February 1969.

That's what heroes are made of — and always will be.

Aquanaut Berry Cannon

(USN Photo)